REFRIGERATION
IN RACE CAR

It is the first book in a large and special series of books, dedicated to motorsport in general; it will cover aerodynamics, suspension, engines, dynamics, etc. ..... Everything you need to learn how to design a full car.

The aim of this series is also to say that I would like to teach again in a university.

I hope that this series will be a success and that I will be able to transmit all my knowledge and all my experience.

@TimoteoBriet

Refrigeration

INTRODUCTION

The principles basics for a good refrigeration, are very simple:

- The equation general for calculate the heat transfer is:

$$Q = U\ A\ \Delta t$$

"Q" is heat; "A" is area of change, "U" is velocity and "t" is difference of temperature.

- If the flow is turbulent, the change of heat, is better:

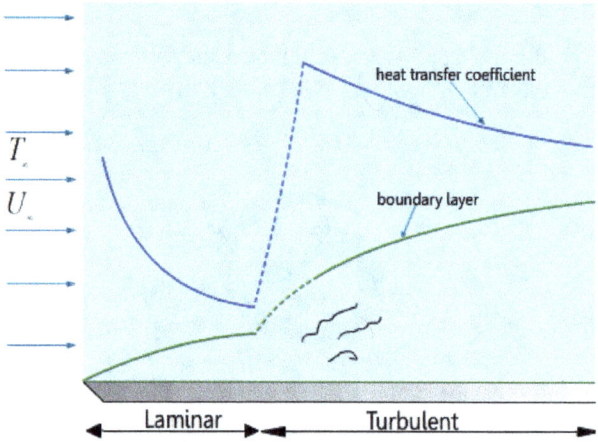

Unfortunately, refrigerate some parts or areas of a car is needed; if it was not strictly necessary, the car could be much more streamlined because the sidepods would not exist or would be radically different.

The parts to be refrigerated are:

- Engine.
- Exhausts.
- Water and oil using radiators.

There are 3 cooling methods:

- Impact.
- Radiator.
- Extraction (suction).

Both the first and second method are based on heat transfer between the material and the air; the third method is a little different, and is based on extracting hot air sucking; To do this flakes are used: the high velocity air passing over creates a vacuum that sucks the air at the end of the duct. These flakes can be placed in many places; "Lovers" work the same way: extracting turbulent air and high pressure air.

At this point, we wanted to mention something important today (2015 season) and due to the cooling requirements bound to make smaller and smaller radiators in order to reduce sidepods and thus increase the maximum speed car, need more dissipation that has doubled the power and energy of the ERS with respect to past years. Every time you brake or accelerate a brutal energy exchange occurs in the system. This exchange cannot be done in any way but in a controlled manner by an electronic power converter. These converters carry some electronic switches that encourage energy exchange. Most of the heat loss occurs in them and that energy is to be dissipated by radiators. Therefore if the loss is reduced by electronic switches radiators size decreases. But how is this achieved? The trick is in the algorithm which controls the switching of the switches. Therefore everything depends on the control software. Control algorithms are infinite and therefore depends primarily on the ability of electronic and control engineers have the equipment. A couple of days ago I heard a researcher who had successfully reduced to 1/10 the size of sinks a power control system for maintaining aircraft system performance. It's amazing.

I guess the F1 teams know all this, at least the big ones. I think this may be one of the keys to the 2014 F1. Renault had heating problems with energy recovery systems and there was talk that they were doing a lot of software changes.

Probably they were working in the control algorithms that are very critical.

TYPES OF FLAKES/ CHIMNEYS

Depending on the geometry of these flakes, you must know the drag they produce: also look at the "Cp" (depression) that result because the less "Cp" more suction force:

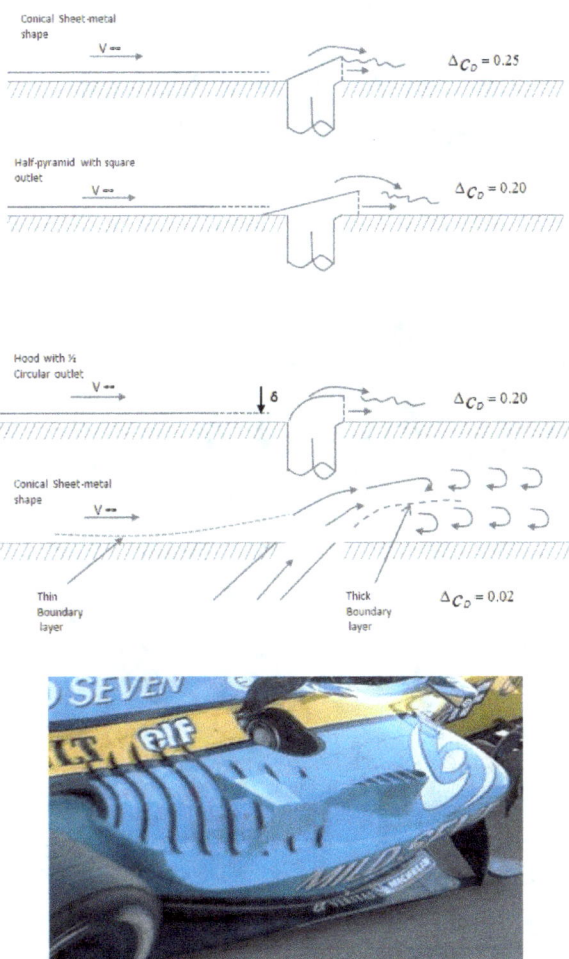

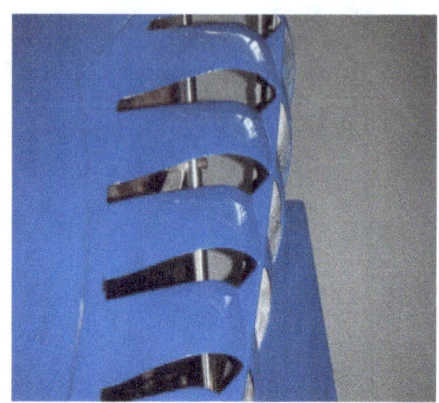

Chimneys work in the same way as the cooling flakes; They can be exploited, due to its shape, for other aerodynamic targets as channel or divert flow.

The same goes for cooling flakes: the extracted air has to go somewhere.

The chimneys are very important because improve the downforce (also augment the drag) and rear wing efficiency. When is open, the flow shifted to rear wing basically, improve his downforce (good flow direction).

Is possible use the shark fin, as a refrigeration help:

The issue of the radiators, mainly due to their large size and therefore its high pressure loss or drag, is an important matter to treat and thereby will be discussed below.

We can make a generalization of the concept of "flakes" making "full flake": in this case it comes to extracting hot air from the engine of the plane:

CREATING LIFT AND DRAG, BY REFRIGERATION

We need radiators; following a fundamental principle of competition, we can place them so that to fulfill its mission, can generate downforce with minimal drag:

$$C_{L_{Cooling}} = -2\frac{V_C A_c A_c}{V_\infty A_f A_0}sin\alpha - \Delta C_{p0}\frac{A_0}{A_f}sin\beta$$

$$C_{D_{Cooling}} = 2\frac{V_C A_c}{V_\infty A_f}\left(1 - \frac{V_C A_c}{V_\infty A_0}cos\propto\right)\frac{A_c}{A_0}sin\alpha - \Delta C_{p0}\frac{A_0}{A_f}cos\beta$$

"A_0" is the departure area; "A_c" radiator area; "A_f" the frontal area of the car; "α" is the angle of exit and "β" the outflow angle.

$$\frac{radiatorspeed}{vehiclespeed} = \frac{V_c}{V} = \frac{1}{1 + \frac{k_p}{4}}$$

$$\frac{V_c}{V_\infty} = \sqrt{\frac{1 - C_{p0}}{\frac{A_c}{A_0}^2 + k_p}}$$

Of course, as we can imagine, if the incident angle is 0 ° radiator heat transfer is much higher (at least in principle); but also the drag originated. The following graphics curious; one must know the transfer depending on the angle of position:

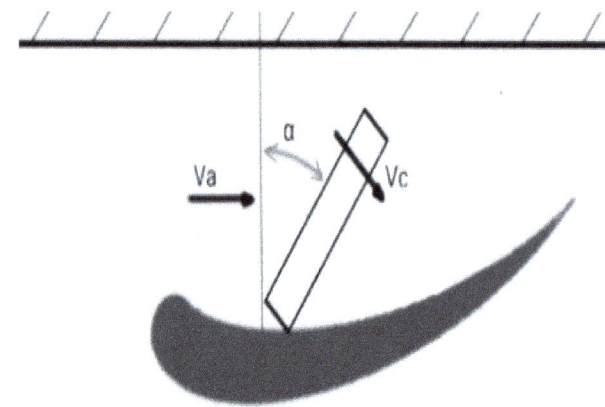

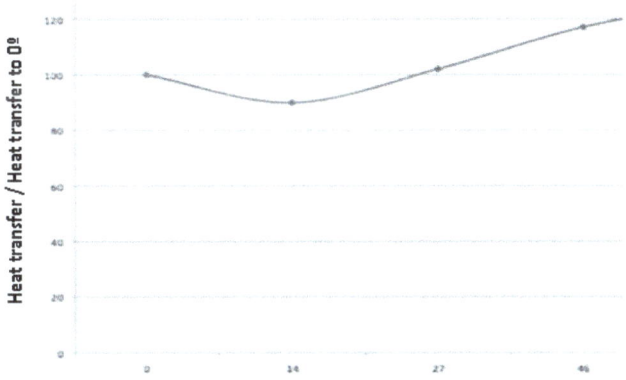

Radiator inclination in degrees

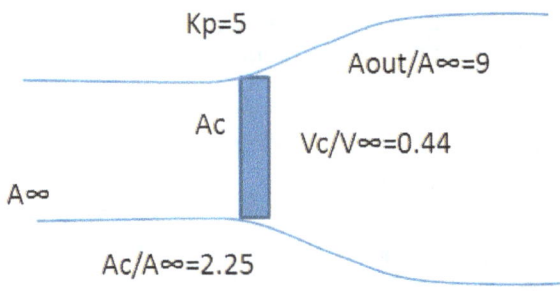

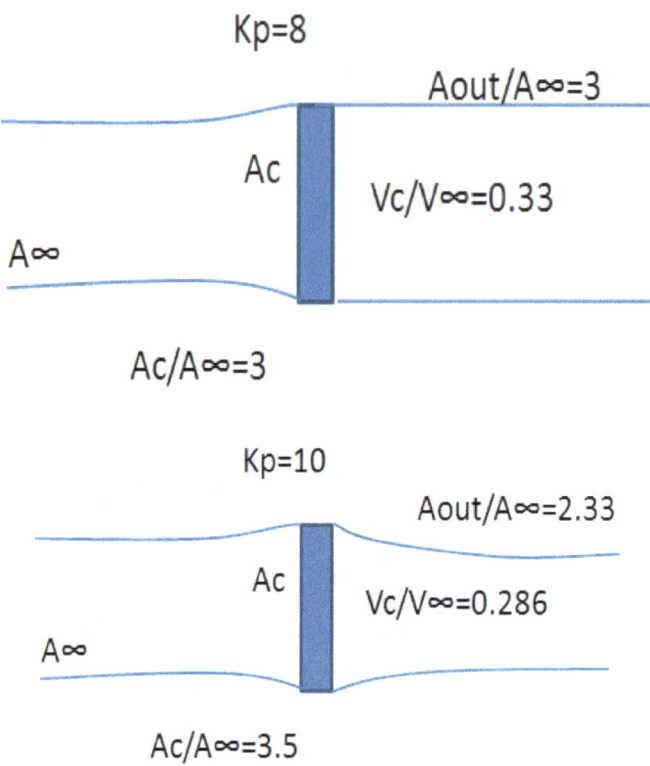

Kp=8

Aout/A∞=3

Ac

Vc/V∞=0.33

A∞

Ac/A∞=3

Kp=10

Aout/A∞=2.33

Ac

Vc/V∞=0.286

A∞

Ac/A∞=3.5

On the other hand, we must know the cooling requirements of each of the elements to be cooled. In the case of a Le Mans category car endurance and speed-dependent: requirements for a GT prototype to 360 km / h (225mph):

Engine water cooler	$6m^3/s$ $(210ft^3/s)$
Engine oil cooler	$3m^3/s$ $(100ft^3/s)$
Gearbox oil	$1m^3/s$ $(35ft^3/s)$

Although we know the needs of flow that has to pass through the duct or the sidepod should aspire, that effectively suck this flow is complicated to know a priori.

We can design the shape of the sidepod so without radiator, the amount of air we need flows, but with the addition of radiator everything changes.

We can do:

- Wind tunnel test: it means having on a reduced scale the car because if it is already built full scale implies change it in case of not properly function (also the radiator to scale, both size and performance).
- CFD test: then we must have not only the CAD model of the car, but we must know the pressure loss of the radiator or similar parameters (losses or tension in each direction); For this we can do 2 things:
  - Wind tunnel test determining the loss of load or resistance.
  - Approximate such loss, in our experience and knowledge.

Both are very complicated and, furthermore, approximated. The tricky thing is to know the size of the radiator, knowing the "factory" characteristics; It is the first and only information we have.

Normally, the design of sidepods or cooling systems are based on the experience of the engineer, though, meet the needs of the flow is a starting data to know. The same applies to the brake cooling, or even more, because just where power brake cooling ais placed is where confluence of several streams and turbulence occur.

The opening of the sidepods and sidepods themselves produce much resistance; and not only that: the radiators themselves also produce a lot of drag: cold days may not need as much cooling because radiators would cool the engine too much being harmful.

Similarly occurs in morning tests; therefore, cover or block the opening of the sidepod using "duct tape"; according to the arrangement of said tape, ie, if it becomes clogged horizontally, vertically, diagonally, and at what rate, it affects in one way or another to the total drag of the car and also the downforce: "CxT" is the total drag, "CzF" is the front downforce, "CzR" is the rear downforce, "Vrad" is the velocity after the radiator and "V0" velocity before radiator:

	$\Delta CxTS$	$\Delta CzFS$	$\Delta CzRs$	$\Delta Vrad / V0$ (cada cara)
No blanking	-	-	-	-
50mm horizontal	0.001	0.002	0.003	-0.6%
100mm horizontal	-0.002	0.001	0.006	-4.1%
150mm horizontal	-0.009	-0.005	0.006	-12.6%
50mm diagonal	0.000	0.000	0.005	-2.4%

Let's see turbulences that can be created downstream:

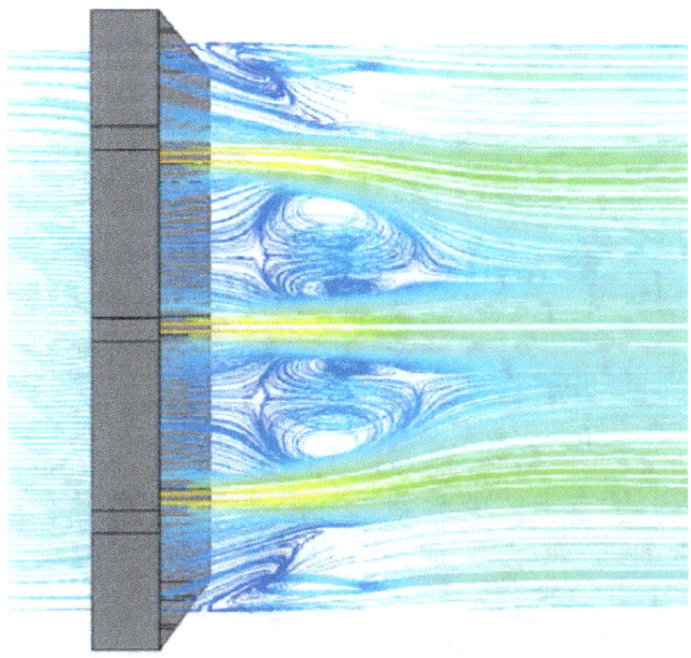

The graphs and data that the manufacturer provides at the buying of the radiator are of the following type:

Radiator	Pressure drop	Dynamic pressure	Speed	Blockage
	Pa	Pa	m/s	.
	1 743,62	186,82	.	9,33
	1 741,13	236,63	17,30	7,36
	1 967,80	151,94	14,30	12,95

It's even possible to get a table of influence regarding the size of the holes of the radiator and with the honeycom or not:

Hole size	Pressure drop	Dynamic pressure	Speed	Blockage
mm	Pa	Pa	m/s	.
3,00	2 052,49	171,87	Fluctuando	11,94
3,00	2 074,91	176,85	14,30	11,73
3,20	1 805,90	261,54	18,00	6,90
3,20	1 793,44	259,05	18,10	6,92
3,40	1 596,66	308,87	20,20	5,17
3,60	1 312,70	378,62	22,10	3,47
3,80	1 028,74	425,94	22,50	2,42

Hole size	Pressure drop	Blockage	Dynamic pressure	Speed
mm	Pa	Pa	m/s	.
3,00	2 074,91	11,74	0,71	14,30
3,00 con Honeycomb	2 002,68	12,97	0,62	14,30
3,80	1 028,74	2,42	1,71	22,50
3,80 con Honeycomb	1 325,15	4,67	1,14	20,90

In any way and form, it is included in a "whole" that is the car itself; study or know their characteristics separately is useless, although some initial values help subsequent definition; once it placed in position and location, we can "really" study efficiency.

Basically:

The diffuser is very important:

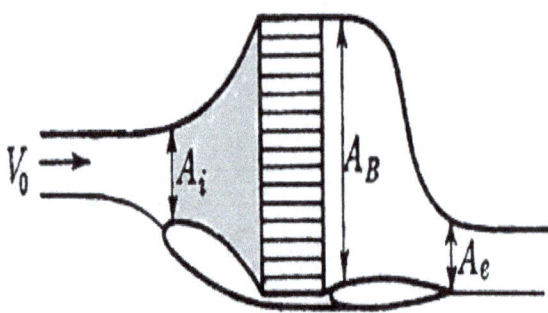

The objective is to have the flow with lower speed so with high pressure; this pressure increases the possibility of traversing the radiator; in fact, this pressure reduce the drag of diffuser because produce one force in walls opposite to drag force.

The geometry diffuser is very important because is necessary not detachment the air in walls; that produce a lot drag.

The zone convergent produces a low pressure and his temperature augment; this zone accelerates the air producing a force forward and expulse the air with heat.

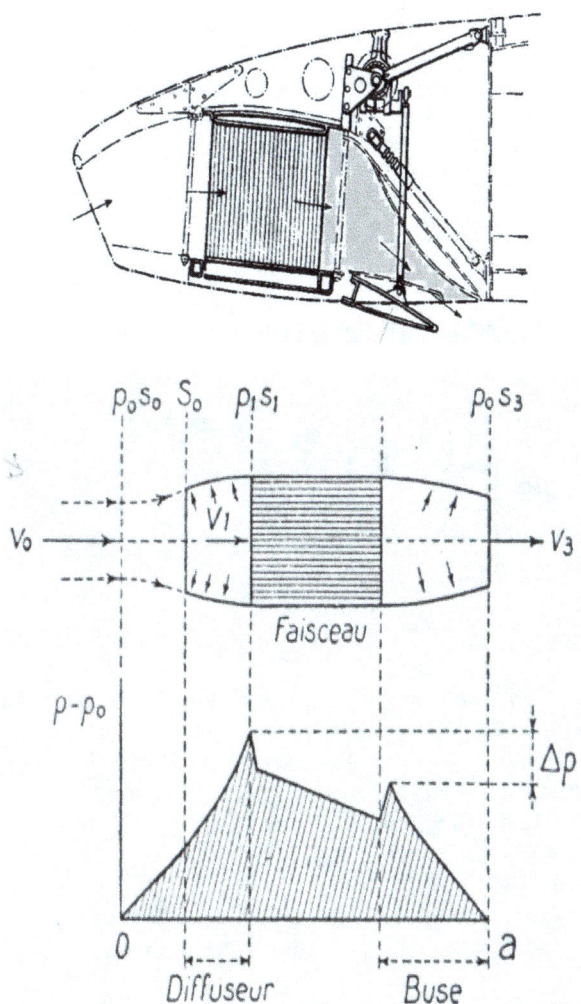

→ Important: if the inlet area is greater than outlet area, may to produce a lot drag. And another think very important: can to generate Hemmoltz vibrations.

SCALED RADIATORS: DUCT CURVE

One of the biggest problems when calibrating a radiator, either for a CFD study or to conduct a scale study of the car in a wind tunnel, is to make a good match in scale, of the most important features defining a radiator. What is intended is to calculate the so-called "duct curve" of the radiator, depending on the porosity or density of the grid. We conducted a test or study using a fan in such a manner we calculate the pressure drop as a function of speed:

$$\Delta P_r = f\,(V_r)$$

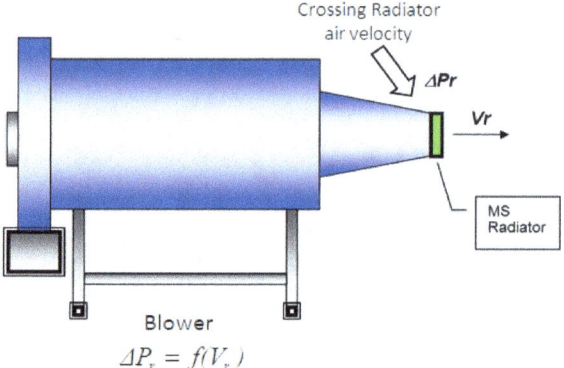

$$\Delta P_r = f(V_r)$$

The manufacturer of the radiator, provides the pressure drop in the "real" car:

$$\Delta P_R = g\,(V_R)$$

We must do a proper similarity between model and reality; to do this, we will use:

$$\Delta Cp = h\,(V\%)$$

ΔP_r Radiator model law:
1. Divide ΔP_r by WT P_{dyn} (wind tunnel).
2. Divide V_r by Wind tunnel speed.

ΔP_R Radiator car law:
1. Divide ΔP_R by P_{dyn} of the car.
2. Divide V_R by car speed.

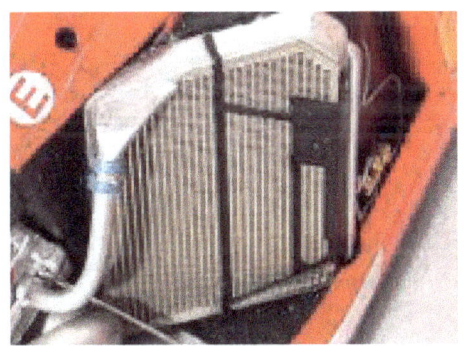

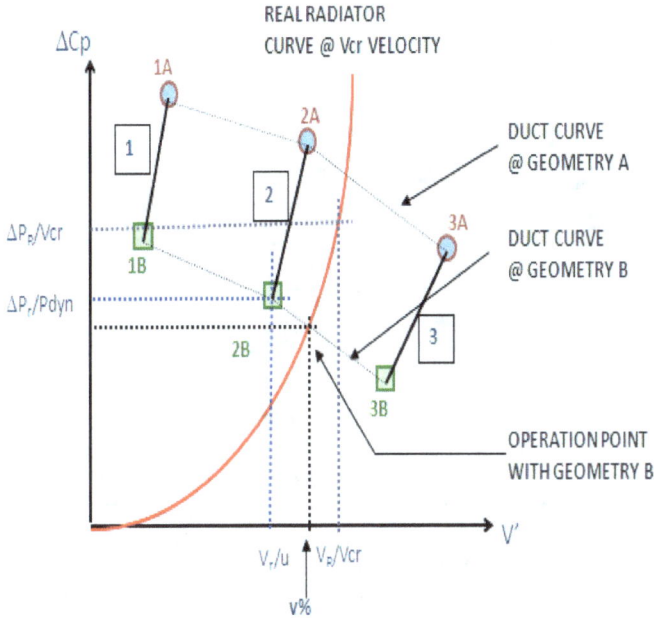

V$_r$: MS radiator crossing velocity for ΔP$_r$
v%: Operation point crossing velocity as per cent of the Vcr
Pdyn: Dynamic pressure in WT
u: WT velocity
Vcr: Reference velocity
ΔP$_R$: Real radiator pressure loss
V$_R$: Real radiator crossing velocity for ΔP$_R$
ΔP$_r$: MS radiator pressure loss during calibration

Duct Curve Example

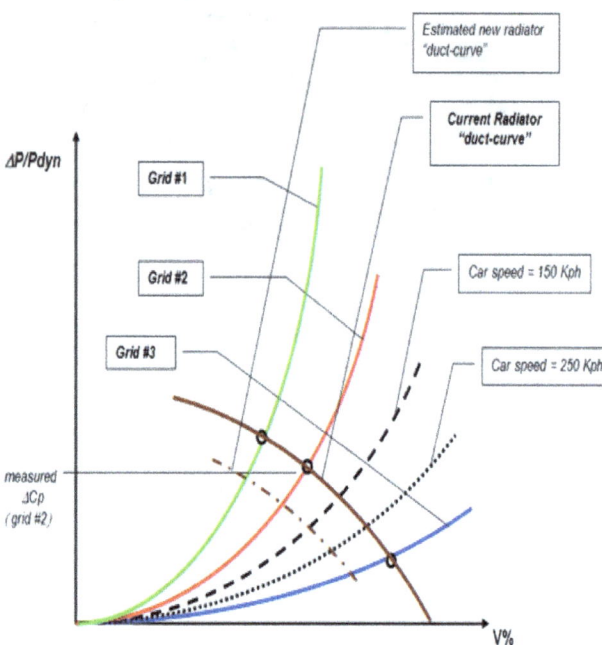

If all this seems complicated to do, keep in mind that the radiator manufacturer do not provide the graphs when the radiator is inside the car. The flux incident on the radiator to study is different in every car; therefore it would be necessary to know the graphic above but having installed the radiator "already" in the car. As you can imagine, it is very difficult and expensive; It is very difficult to know something "real" from something "imaginary" or virtual; normally iterating (testing and measurement) gets closer and closer to the real solution without ever reaching.

All the flux reaching the radiator depends on the form of the whole car.

NACA INTAKES

Another important element cooling is called "Naca intakes". They are openings in the surface of the car that "collect" air and lead it where required; these shots have the "essential" characteristic of producing very little drag; they work as follows:

They have very sharp edges that produce turbulence or vortex inwards; then creates a vacuum that sucks the flow into the duct opening; non-differentiable corners are not liked by the Navier Stokess; but in this case, they must be necessarily pointy; it is one of the few cases:

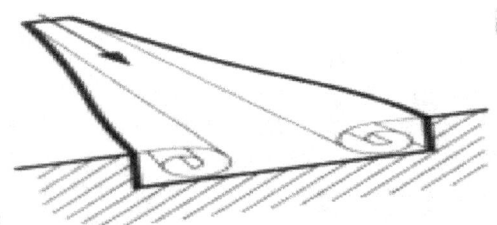

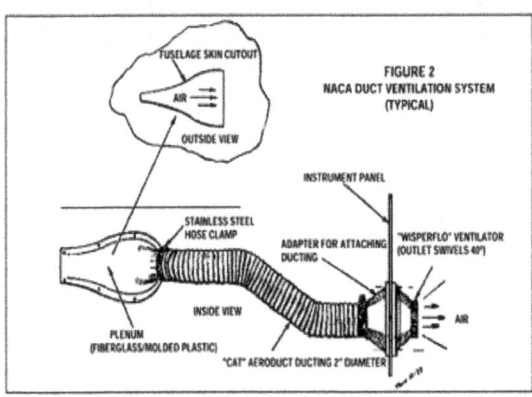

Naca intakes are basically used for three objectives:

- Cooling the driver.
- Cooling the brakes.
- As intake port for the engine.

Let's see two pictures of Mosler MT900R with brake cooling intakes and admission, composed of Naca intakes:

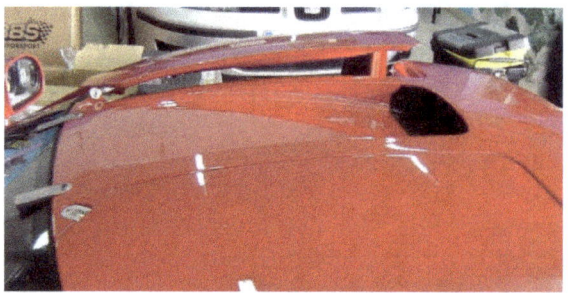

Some designing rules of Naca intakes:

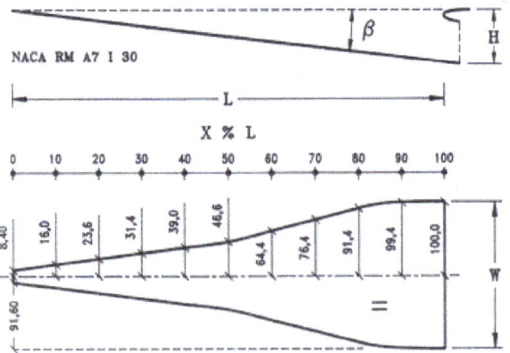

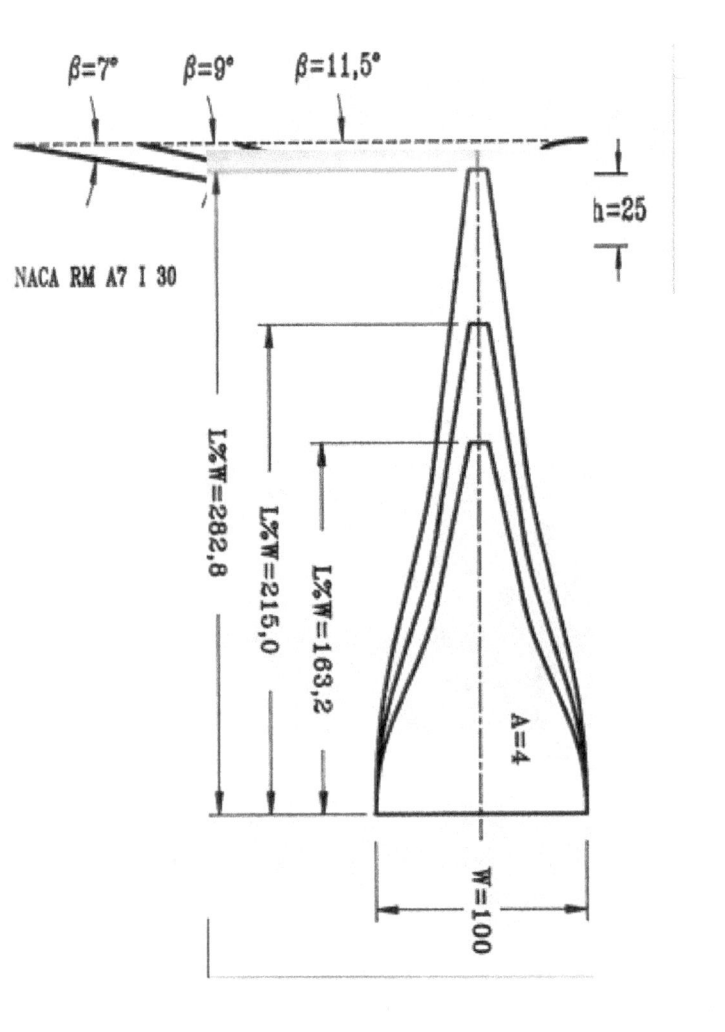

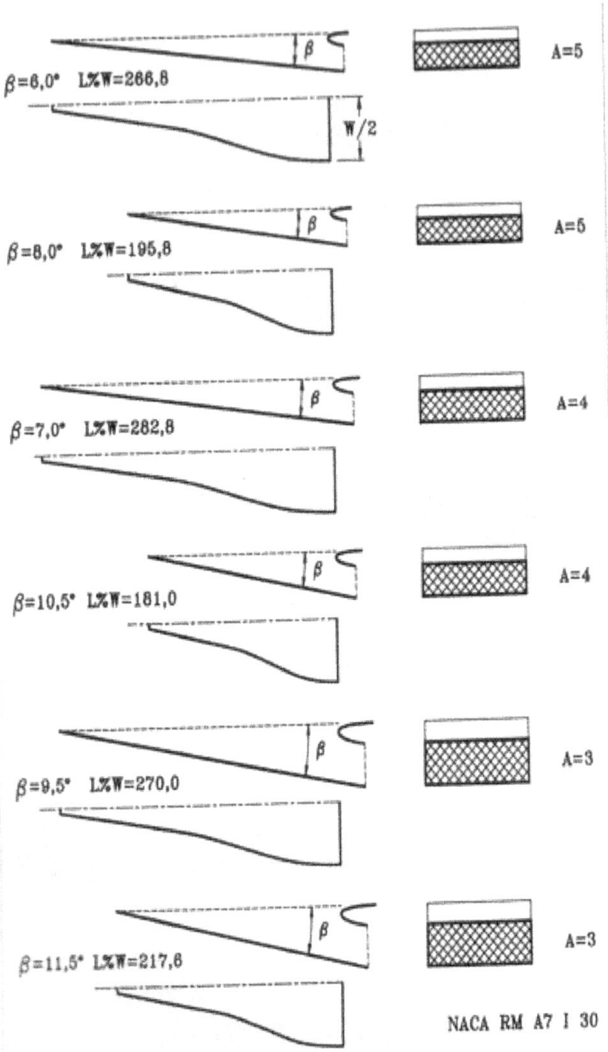
β=6,0° L%W=266,8
β
W/2
A=5
β=8,0° L%W=195,8
β
A=5
β=7,0° L%W=282,8
β
A=4
β=10,5° L%W=181,0
β
A=4
β=9,5° L%W=270,0
β
A=3
β=11,5° L%W=217,6
β
A=3

NACA RM A7 I 30

BARGE BOARD

At the front of the sidepods, there are some pieces called "bargeboard"; its function is twofold and both very important:

Cause a drop in the speed behind, right at the entrance of the sidepods, to improve cooling (as a diffuser).

Make the surrounding air does not enter the floor of the car (or penetrate properly), increasing the downforce.

This second function, perhaps the most important because the car floor and diffuser, produces approximately 75% of the total car downforce.

These deflectors produce a high energy vortex which can aerodynamically seal the floor.

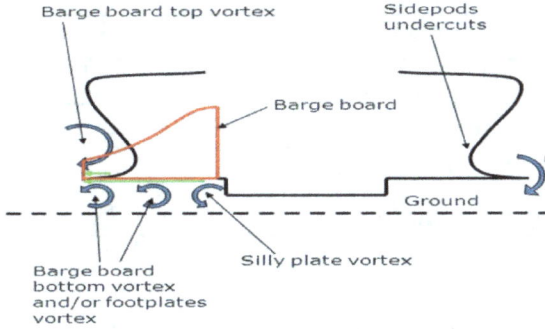

We can make a "real" small test to know how, where and to what extent the bargeboards influence the aerodynamics of the car; to do this, we remove them and see the variations in percentages:

% Change in forces, without bargeboards

Refriferation	7
Aero eficiency	-3
Aero balance	-1,5
Total drag	-1
Drag body	-1,9
Rear downforce	-2
Front downforce	-6
Body downforce	-3,9

Moreover, the influence of the bargeboards in different aerodynamic parts of the car may involve an improvement in downforce of all of them:

% Change in forces, with bargeboards

Rear wing	2,4
Front wing drag	0,4
Rear wing downforce	1,9
Front wing downforce	0,4
Wheel rear drag	0,45
Wheel front drag	0,8

The undercut sidepode (zone), have a differents functions:

- Help to Hammer head to add downforce.
- Seal the floor, so produce a lot downforce.
- Help also to barge board, producing downforce.

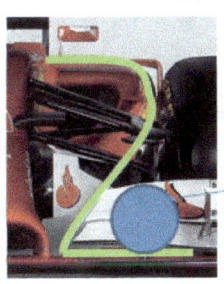 

Undercut

The sidepode leading edge flickups, help to hammer head function; basically is that. But produce a lot drag; so is important to improve this element.

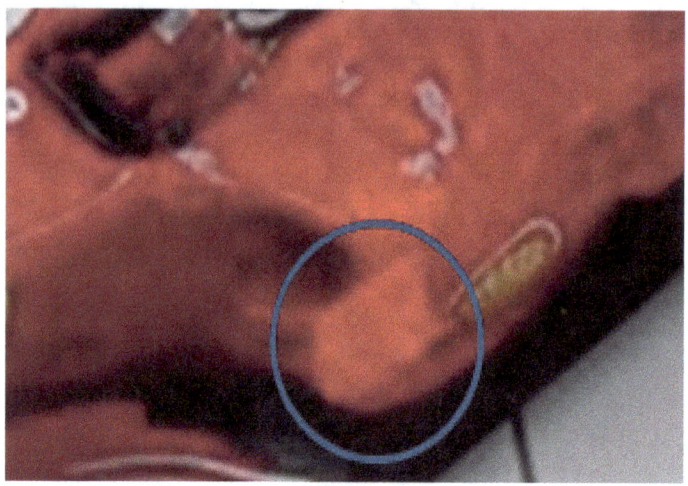

Hammer head

BRAKE COOLING

We know the complexity that any attempt to design a cooling system "properly" and that works perfectly.

Regarding the cooling of the entire braking system, it comes cooling the brake disk, the caliper and the entire system that supports and holds these elements; to do this one of the most used systems consists in 2 intakes one inside the other, normally:

The outer one supplies air flow to a side of the disc and the brake caliper, while the other, the inner one, provides air flow to the center of the disc and via holes and channels performed inside the disc itself inside channels cools inwardly.

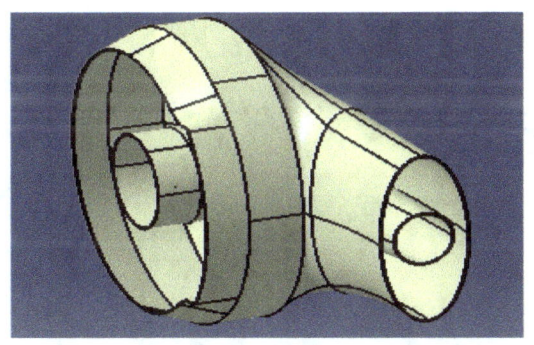

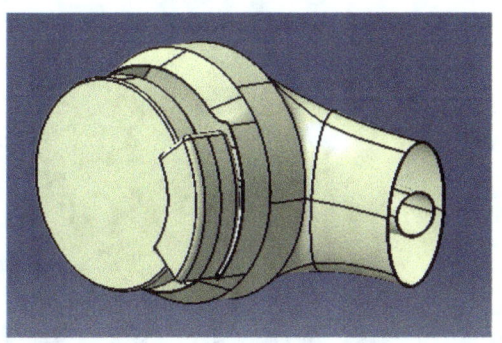

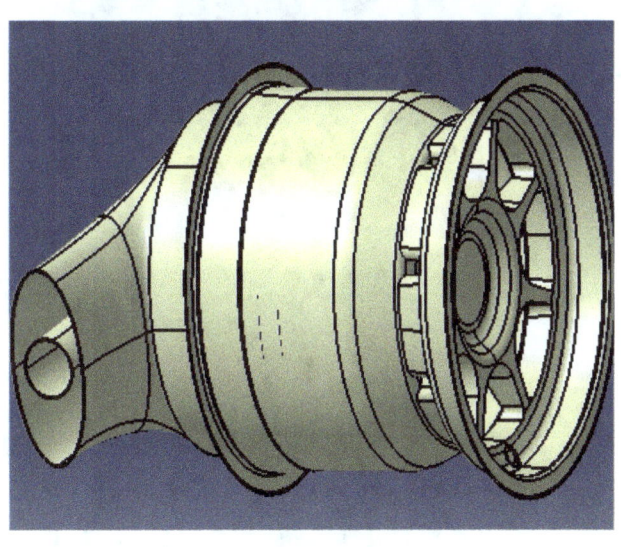

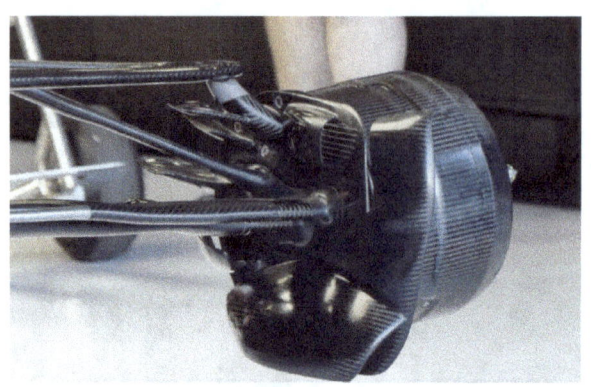

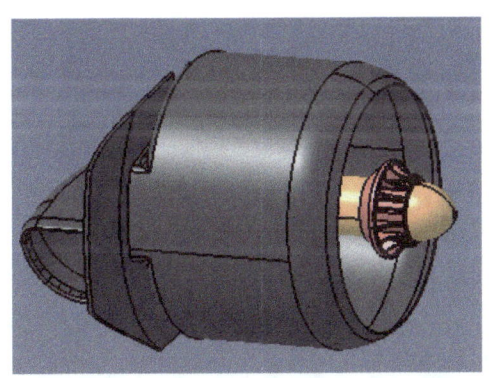

Other method for improving rim refrigeration, is try to augment the surface in interior rim. That is:

The brake cooling ducts can be varied because should be placed in many different positions: with air damm, splitter, open or closed wheels, etc.

And as always, taking the "necessary" brake cooling, can use it to "other things", eg. Heating the tires is essential for teams that have problems reaching the tire optimum temperature, what is intended is to use the hot air to heat the rim, or directly the tire.

Otherwise it, it is conveniently extracted in order to not overheat the tires.

About the brakes discs, there are a lot holes in order to extract the air:

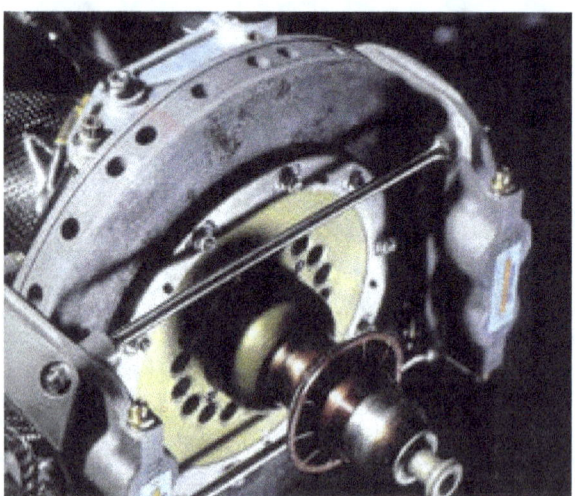

Is possible to change the geometry of ending edge, in order to extract the air in other way more efficient:

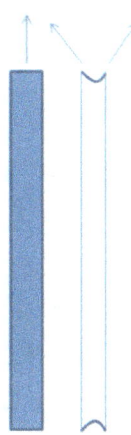

POSITIONING THE RADIATORS

Why aren't the radiators placed perpendicular to the air flow?

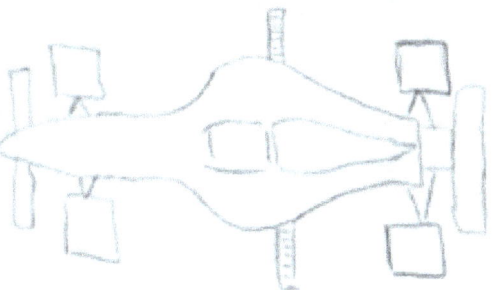

It consists of alveoli whose maximum depth is 10 cm; suppose maximum speed of 75 m / s:

The thickness of the laminar boundary layer is $0.472/(Re)^{0.5} = 0.668$ mm.

Therefore:

- The heat exchange layer is too thin for the heat transfer between the radiator and the air
- The passage time is too short: $0.1/75 = 1.3 / 100$ seconds.
- The rate of airflow is too fast.
- If the incidence of flow varies, the alveoli cooling function poorly.

Radiators position, depending on their number or quantity:

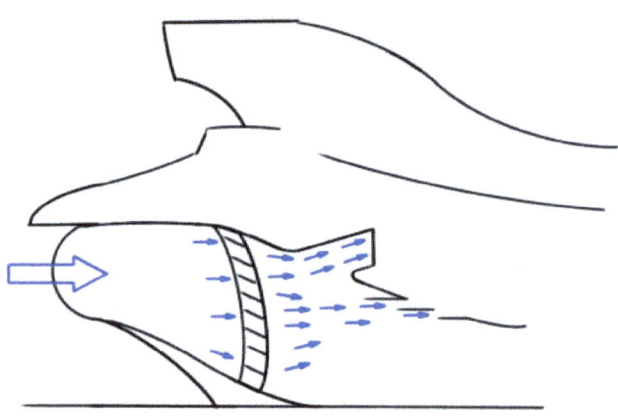

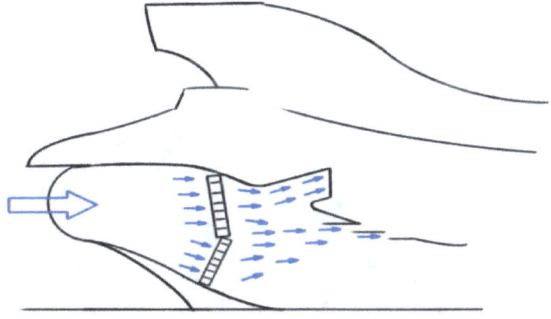

In case of three radiators:

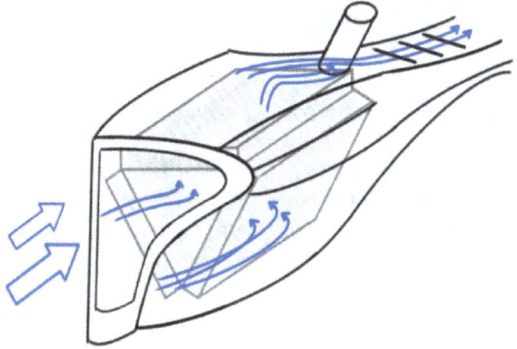

Normally:

Let us suppose extremely dirty circuits; refrigeration would be compromised if elements covering radiators themselves or air inlet were introduced.

For this, a series of grids are placed in the inlet of the sidepods or inside them, with a very particular geometry (plant view):

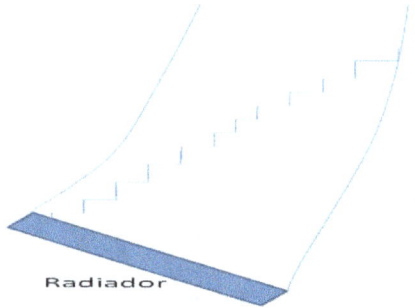

Radiador

It is a grid whose vertical walls are inclined approximately 5º:

5º

This special geometry, would allow enter a paper, plastic or rubber, and cover only the horizontal walls (according drawing) still allowing the passage of airflow through the vertical or inclined walls.

INSTALLATION AND DESIGN/ DIMENSIONING RADIATORS

INSTALLATION OF RADIATORS

An important issue in the installation of a radiator is the efficiency of the radiator; that is, if "all" the radiator functions like one or if there is somewhere where is malfunctioning.

To do this, we can do two things:

- Using CFD, or by using infrared cameras or temperature sensors, "see" if the radiator and its entire surface, acts as such.
- Rely on the following installation premises.

The biggest problem that is had is the pressure drop and flow separation at the ends; This causes the flow on the extreme part of the face of the radiator be turbulent and make that part of the radiator does not work properly:

Assume the following sidepod design and layout of radiator:

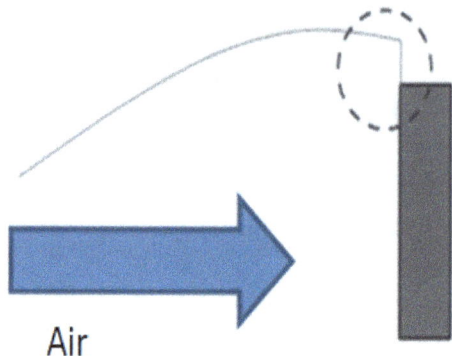

Air

In the marked area, there is a pressure drop and flow separation, detrimental to the radiator and its functioning; It is possible to change geometry at this point to improve efficiency.

Another issue is the outflow of air just in the radiator; by experience, it is advisable to practice an extension of the section of the radiator to mitigate losses and therefore increase performance; with these two considerations, the design is now:

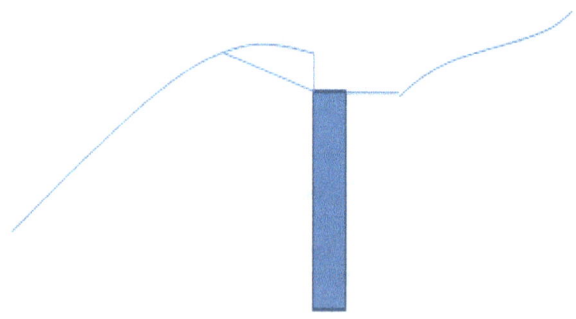

Another possible system to use, is the next:

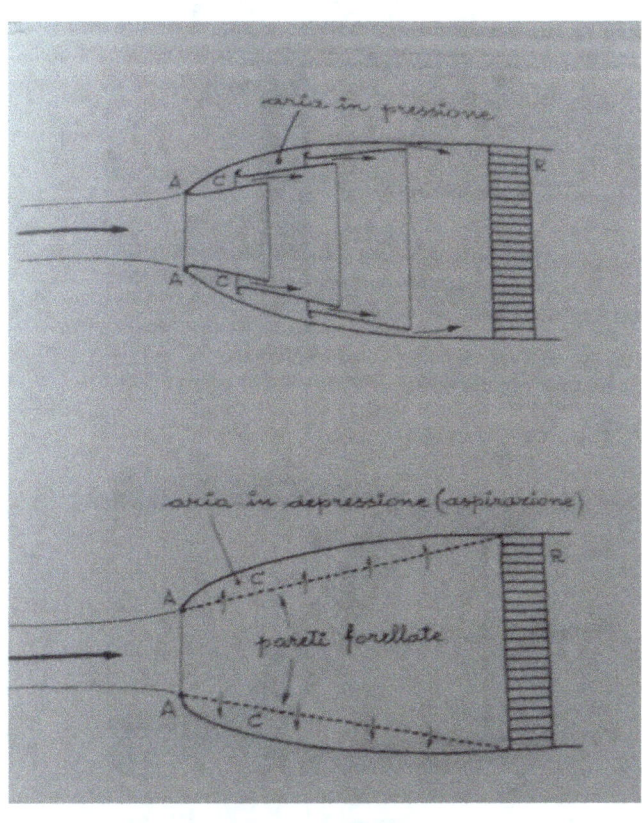

DESIGNING AND DIMENSIONING RADIATORS

Show below, a system or procedure tested by Enrique Scalabroni:

TYPES LAYOUTS REFREIERATION SYSTEMS
TYPE 1

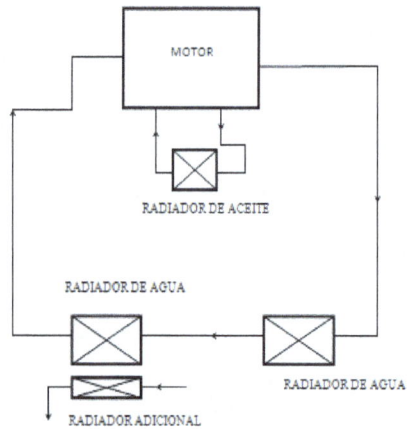

Two water radiators for cooling engine have, which have a cross flow. The engine oil is cooled by another radiator through which air circulates. Addition can have an independent additional radiator system, which is responsible for cooling transmission oil.

TYPE 2

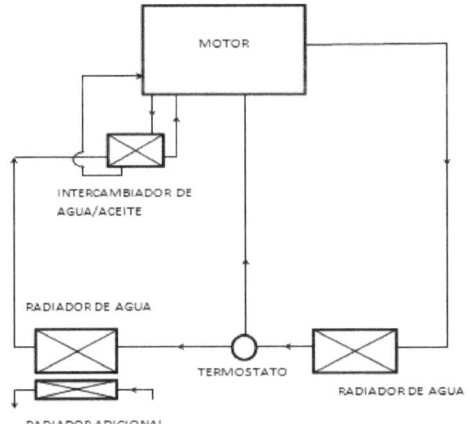

Two water radiators connected to a thermostat according to the temperature thereof directs the flow to the motor or to the second radiator with water. This system also has a heat exchanger water / oil (intercooler), with which engine oil is cooled with water (without mixing both). You can have an additional radiator for cooling the oil gearbox, independent of the previous system.

TYPE 3

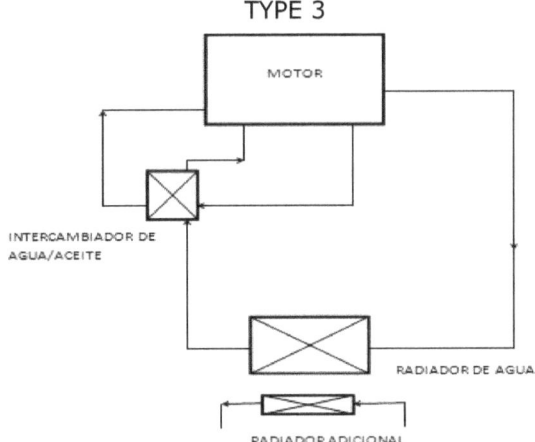

The water flowing through the motor is cooled by radiator water, whereby through exchanger water / oil (intercooler) this same water for cooling the engine oil is used (without mixing both). You can have an additional, independent radiator system for cooling oil gearbox.

REFRIGERATION CALCULUS EXPRESSIONS:

- [ ] Calor rechazado por el sistema de agua:
$$Q_{(water)} = [kW]$$

- [ ] Calor rechazado por el sistema de aceite:
$$Q_{(oil)} = [kW]$$

- [ ] Flujo de masa de la bomba de agua:
$$\dot{m} = \left[Kg/min \right]$$

- [ ] Calor especifico del agua a presión constante:
$$C_p = \left[kJ/(kg\,^0C) \right]$$

- [ ] Temperatura del agua saliendo del motor y entrando al radiador: $T_{water(IN)} = [^0C$

- [ ] Temperatura del aire ambiente:
$$T_{air(IN)} = [^0C$$

- [ ] Temperatura del aceite a la salida del motor y entrada del radiador: $T_{oil(IN)} = [^0C$

- [ ] Diferencia extrema de temperatura: $ETD = [^0C$

- [ ] Diferencia de caída de temperatura:
$$\Delta T = [^0C$$

- [ ] Eficiencia del ducto: ξ_D

- [ ] Rata de expansión del ducto de admisión: ρ_E

- [ ] Velocidad del automóvil: $V_C = \left[\dfrac{km}{h}\right]$

- [ ] Velocidad de cara: $V_F = \left[\dfrac{m}{s}\right]$

- [ ] Área del radiador: $A_R = [m^2]$

- [ ] Área del ducto de admisión de aire:
  $A_{(i\Delta)} = [m^2]$

- [ ] Primera área del radiador de agua:
  $A_{R(1)} = [m^2]$

- [ ] Segunda área del radiador de agua:
  $A_{R(2)} = [m^2]$

- [ ] Área aceite/aire del radiador: $A_{R(3)} = [m^2]$

- [ ] Área del intercambiador de calor (Intercooler)
  agua/aceite: $I_{(w/o)(4)} = [m^2]$

- [ ] Área adicional adentro del radiador principal:
  $A_{R(5)} = [m^2]$

- [ ] Área total del radiador principal:
  $A_{R(MT)} = [m^2]$

- [ ] Área total de radiador: $A_{(T)} = [m^2]$

- [ ] Área efectiva del radiador: $A_{(E)} = [m^2]$

☐ Efectividad de la rata de expansión del ducto de entrada: $\rho_{E(E)}$

$$V_{critica} = \frac{V_F}{\rho_{E(E)}}$$

☐ Velocidad critica del automóvil:

$$= \left[\frac{m}{s}\right]$$

EQUATIONS:

☐ $Q_{(total)} = Q_{(water)} + Q_{(oil)} = [kW]$

☐ $$\Delta T = \frac{Q_{(total)}}{\dot{m} * C_p} = [^0C$$

☐ $$ETD = \begin{Bmatrix} T_{water(IN)} - T_{air(IN)} \\ T_{oil(IN)} - T_{air(IN)} \\ T_{oil(IN)} - T_{water(OUT)} \end{Bmatrix} = [^0C$$

☐ $$\frac{Q}{ETD} = \frac{Q_{(total)}}{ETD} = \left[\frac{kW}{(^0C}\right)$$

☐ $$V_F = VC * \rho_E = \left[\frac{m}{s}\right]$$

☐ $$A_R = \frac{\dfrac{Q}{ETD}}{\left(\dfrac{Q}{ETD} * \dfrac{1}{Area}\right) * \xi_D} = [m^2]$$

☐ $A_{R(MT)} = A_{R(1)} + A_{R(2)} = [m^2]$

☐ $A_{(T)} = A_{R(MT)} + A_{R(5)} = [m^2]$

$$\square \quad A_{(E)} = A_{(T)} * \xi_D = [m^2]$$

$$\square \quad \rho_{E(E)} = \frac{A_{(iD)}}{A_{(E)}}$$

$$\square \quad V_F = Vc * \rho_{E(E)} = \left[\frac{m}{s}\right]$$

$$\square \quad \frac{Q}{ETD} = A_{R(MT)} * \xi_D * \left(\frac{Q}{ETD} * \frac{1}{Area}\right)$$

$$\square \quad ETD = \frac{Q_{(total)}}{\dfrac{Q}{ETD}}$$

$$\square \quad T_{fluid(IN)} = ETD + \left\{ \begin{array}{c} T_{air(IN)} \\ T_{water(OUT)} \end{array} \right\} = [^0C$$

$$\square \quad V_{critica} = \frac{V_F}{\rho_{E(E)}}$$

CALCULUS PROCEDURE:

- Knowing the type of layout of radiators used as well as the type of heater use.

- The data must meet before being introduced to the program in addition to the design data; They are the following: average vehicle speed and average temperature on that track, maximum power, pump flow of water, percentage of heat dissipated by the water or coolant, heat dissipated by the oil rpm engine specific heat constant fluid pressure, fluid density (water or refrigerant), the temperature of the fluid input and output efficiency of the pipeline expansion coefficient inlet duct and define the amount of heat to be dissipated by the first radiator.

- Data track (track speed and temperature) are introduced.

- Maximum power is introduced, pump flow, percentage of heat dissipated by the coolant or water, heat dissipated by the oil, engine rpm.

- With these data's we are given the result of the total heat to dissipate, and then decide how you want to dissipate heat to the first (or only) radiator.

- The type of main radiator is chosen.

- After enter data on the part of the main radiator, such as: Specific heat of the fluid at constant pressure, density of the fluid (water or refrigerant), fluid temperature at the input and output efficiency of the pipeline expansion coefficient inlet duct and the amount of heat to be dissipated.

- Enter data in the second radiator, which are: fluid temperature at the entrance, pipeline efficiency, coefficient of expansion of the inlet duct.

NOTES:

The heat dissipated by the only first) radiator may not be greater than the total heat dissipated.

The speed of the car to take is the same average speed on that track.

The ambient temperature is the average ambient track temperature, measured one meter above the ground.

If the radiator not be chosen within the program, choose the type of custom radiator and search the same graphical value $\dfrac{Q}{ETD} * \dfrac{1}{Area}$; knowing ΔT y V_F.

+ Notes:

- Get a greater number of curves of other types of radiators, to be annexed to the program, to simplify time in obtaining results and give the user selection options.

- Cooling is an indisputable and necessary fact; if they could do the smaller pontoons, the car's top speed increase because there would be less drag.

I imagine a lower lateral outlets or cooling to reduce the need for such large sidepods....

➜ For calculating the porosity or permeability factor:

$$q = \rho \frac{V^2}{2}$$

$$C = \frac{\Delta P}{q}$$

$$\lambda = \frac{1}{\sqrt{1+C}}$$

"λ" is
pressure (botl
radiator and "ρ

CALCULUS PRC

REPORT

Ikuzawa International Ltd.

Number   KC0504

Subject   Cooling performance prediction

Date   7 October 1994

Author   K Cashmore

Introduction

This paper is based on work originally done by Dr R D Williamson.

In order to estimate the heat rejection capabilities of any car cooling set-up we must be able to predict the mass flow through the radiators.

We measure the pressure drop across the model radiators in the wind tunnel. If we have data for pressure drop Δp versus core velocity v for the model core then we can look up the velocity from the pressure drop. The model core Δp versus v characteristic is not linear with v (that is k_{core} is constant with varying v). It is highly Reynolds number dependant. Also it varies in a different fashion to that of the full size radiator core which means that we cannot scale up the model core velocity to predict the car core velocity.

In general we can match k_{core} for the model and the car at only one speed ratio

$$VR = v_{car} / v_{freestream} = v_{model} / v_{freestream}$$

We can therefore only directly project model core mass flow to the car core mass flow at this condition. This is a problem given the need to estimate cooling at different car speeds and the fact that different model configurations result in different model core velocities, altering VR.

A further problem is that if the car core permeability is altered (a different matrix is used) then a new model core must be used and the new results will not be comparable to the old.

One solution to this is to take an energy based approach, identifying and quantifying losses which can be scaled from model to car (i.e. proportional to v^2) and those which cannot.

A general method is created whereby model results may be mapped more accurately to full size results, and furthermore model results gathered with cores of differing permeability may be compared to each other and mapped to car radiators of differing permeability.

Theory

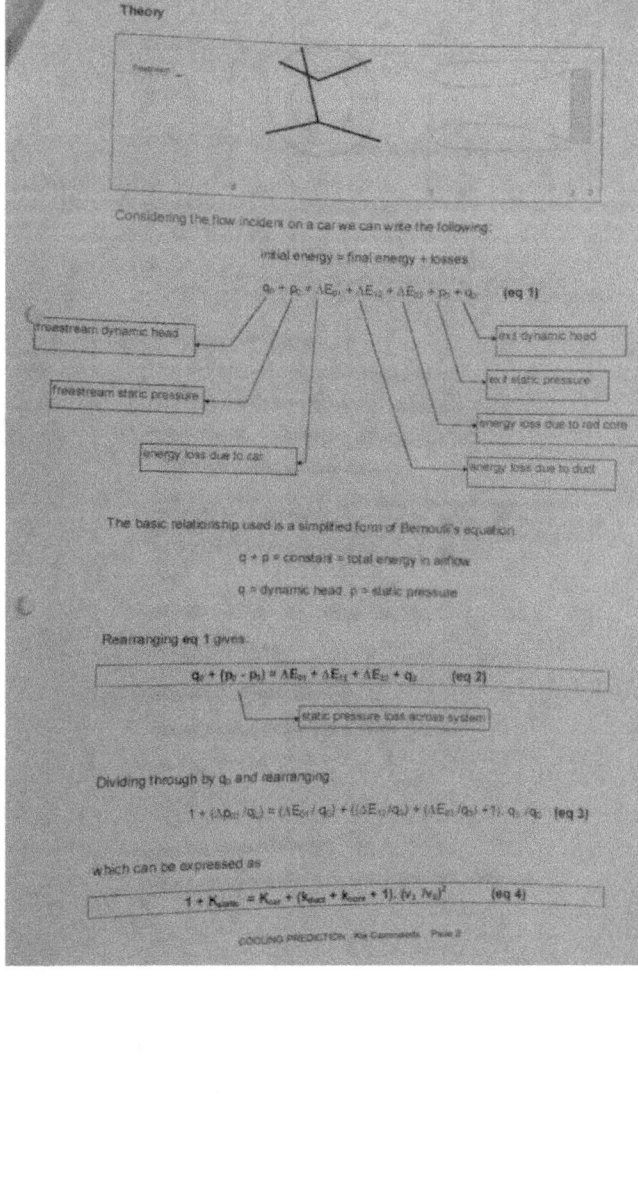

Considering the flow incident on a car we can write the following:

intial energy = final energy + losses

$$q_0 + p_0 + \Delta E_{gr} + \Delta E_{rc} + \Delta E_{cd} + p_2 + q_2 \quad \text{(eq 1)}$$

freestream dynamic head

freestream static pressure

energy loss due to car

exit dynamic head

exit static pressure

energy loss due to rad core

energy loss due to duct

The basic relationship used is a simplified form of Bernoulli's equation.

$$q + p = \text{constant} = \text{total energy in airflow}$$

$$q = \text{dynamic head} \quad p = \text{static pressure}$$

Rearranging eq 1 gives:

$$q_0 + (p_0 - p_2) = \Delta E_{gr} + \Delta E_{rc} + \Delta E_{cd} + q_2 \quad \text{(eq 2)}$$

static pressure loss across system

Dividing through by q_0 and rearranging:

$$1 + (\Delta p_{st} / q_0) = (\Delta E_{gr} / q_0) + ((\Delta E_{rc}/q_0) + (\Delta E_{cd} / q_0) + 1) . q_2 / q_0 \quad \text{(eq 3)}$$

which can be expressed as

$$1 + K_{static} = K_{car} + (k_{duct} + k_{core} + 1) . (v_2 / v_0)^2 \quad \text{(eq 4)}$$

where

K = loss coefficient referred to freestream dynamic head q_1

k = loss coefficient referred to core dynamic head q_2

For the model:

Note that p_0, p_1, p_2, q_0 and $(p_1)_{loss}$ must all be measured directly in the tunnel.

From eq 2

q_1 = freestream dynamic head - measured in tunnel

$q_2 = \frac{1}{2} \rho v_{core}^2$ = core dynamic head = q_{core}

> This cannot be measured directly but can be inferred from previously measured data of pressure drop versus core velocity for the particular model core in use

$p_1 - p_2$ = overall static pressure drop

> p_1 = freestream static pressure - measured in tunnel

> p_2 = radiator back face static pressure - measured in tunnel

ΔE_{01} = energy loss due to car (front wing, tyre, suspension etc.)

> = freestream energy - energy at inlet entry

> This can be measured directly by blanking off the model radiator and measuring the static pressure at front face (this assumes that there will be no resultant change to the flow field around the car and that there is no back pressure effect on the pressure drop across the radiator)

> Then

> $\Delta E_{01} = (p_1 + q_1) - (p_1 + q_1)$ but $(q_1 = 0)_{loss}$ so

> $\Delta E_{01} = p_1 + q_1 - (p_1)_{loss}$

ΔE_{12} = energy loss due to duct inefficiency

> = energy at inlet to duct - energy at radiator front face

> = $(p_1 + q_1) - (p_2 + q_2)$

> = $(p_1)_{loss} - p_2 - q_{core}$

ΔE_{23} = energy loss across radiator core

> As q is the same both sides of the core (conservation of mass) then

CALCULATING PARAMETERS RADIATORS

In a lot softwares applications, there are some radiators parameters, as a Inertia drag, etc... for calculate these values, is necessary to work as that:

l = Length of the fin.
μ = Viscosity.
Δp = Pressure drop.
V = Velocity.
C_2 = Inertial resistance factor.
S = Sourde term.
ρ = Density.
A = Permeability.

$$\frac{\Delta p}{l} = S = -\left(\frac{\mu}{\alpha}V + C_2 \frac{1}{2}\rho V^2\right)$$

This procedure is possible to do it by CFD or Wind Tunnel.

$$\frac{\Delta P}{l} = a\,V + b\,V^2$$

$$Inertial\ coefficient = \frac{2\,b}{\rho\,l}$$

$$Viscous\ coefficient = \frac{a}{l\,\mu}$$

We can see one sample about:

We simulate one element of radiator (one fin of 38 mm into wind tunnel virtual of 238 mm length):

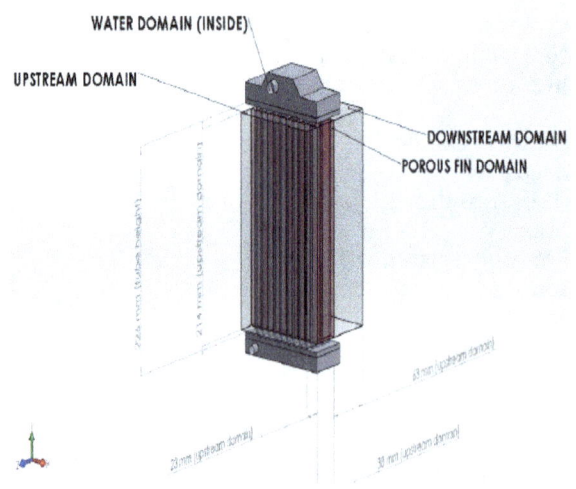

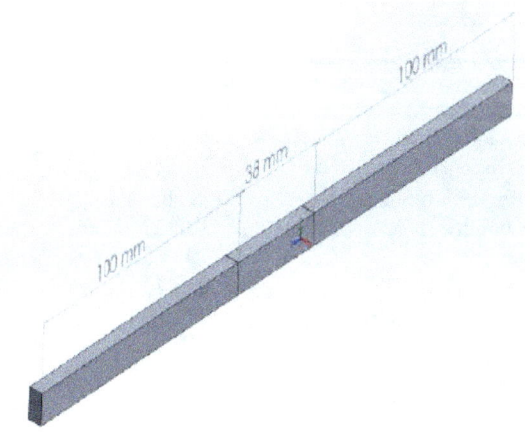

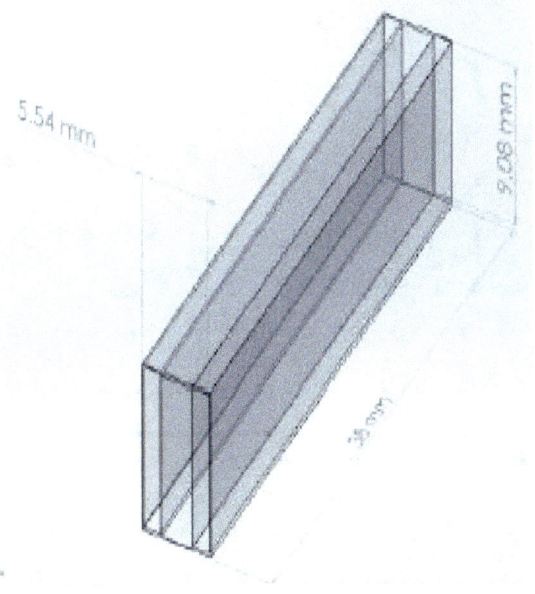

Input Parameters for Unit Cell Straight Fin Simulations

	DESCRIPTION	Unit
Domain length	38	mm
Element number	3,624,060	
Skewness (average)	0.22	
Turbulence modeling	k-ε-realizable	
Fin volume	108.07	mm^3
Total volume	1936.8	mm^3
Porosity	0.9442	
Hydraulic diameter	0.00241	
Turbulence Intensity	0.053	
Turbulence length	0.000169	
Solution method	SIMPLE	
Computation time	11	mins

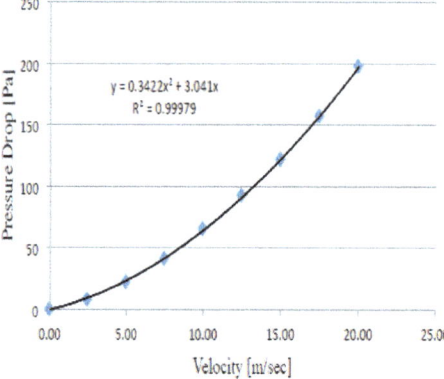

Heat transfer characteristics for a unit cell of a straight fin

Interfacial area	Porous volume	IAD	HTC	T_{ref}
(m^2)	(m^3)	(1/m)	$(W/m^2\text{-}K)$	(K)
0.001567621	1.93678×10^{-6}	809	133	321

Porous jump coefficients for a unit cell of a straight fin

	Face permeability $(1/m^2)$	Thickness (m)	Inertial coefficient $(1/m)$
Inlet	4.49×10^6	0.1	1.54
Outlet	4.49×10^6	0.1	-3.6

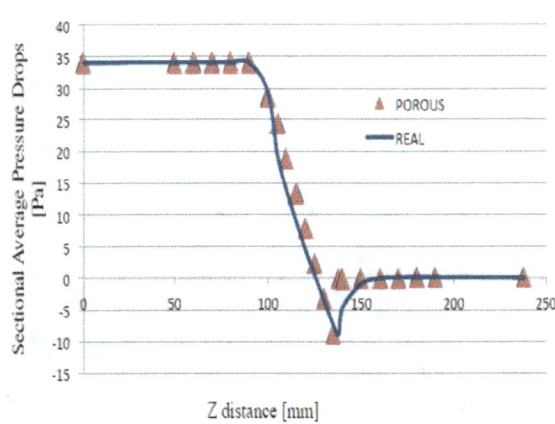

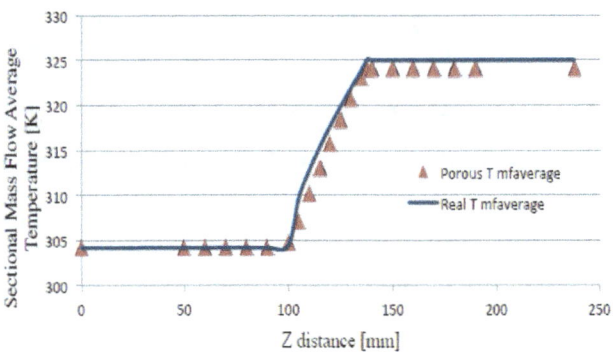

Is possible also to calculi that, from test "real":

RADIATOR PANNEL 'K' FACTOR:

$$K = K_p$$

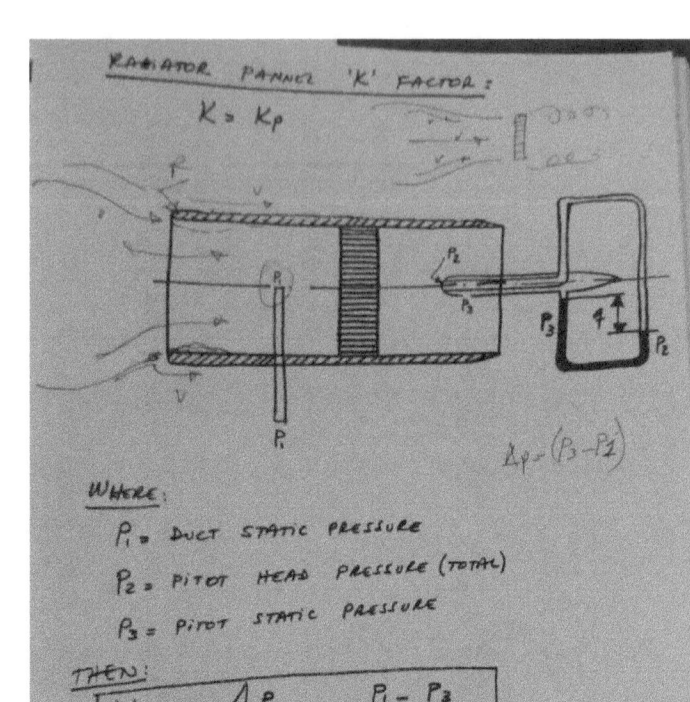

$$\Delta p = (P_3 - P_1)$$

WHERE:

$P_1 =$ DUCT STATIC PRESSURE

$P_2 =$ PITOT HEAD PRESSURE (TOTAL)

$P_3 =$ PITOT STATIC PRESSURE

THEN:

$$K = \frac{\Delta P}{q} = \frac{P_1 - P_3}{P_2 - P_3}$$

$q =$ DYNAMIC PRESSURE $= \frac{1}{2} \cdot \rho \cdot V^2$

NOTE: MANOMETER HEIGHTS MAY USED SINCE
K IS NON-DIMENSIONAL

SPEED IN RADIATOR FACE:

$$\boxed{V_1} = \frac{A_0 \cdot V_0}{A_1} = \frac{(.0628) \cdot (83.33)}{(.144)} = \boxed{36.34 \ [m/s]}$$

$$V_1 = 130.82 \ [Km/h]$$

PRESSURE RADIATOR FACE:

$$p_1 = 10\,448 + \frac{1}{2} (.125) \cdot \left\{ (83.33)^2 - (36.34)^2 \right\} =$$

$$\boxed{p_1} = 10\,448 + 351.46 = \boxed{10\,780 \ [kg/m^2]}$$

INCREMENT IN PRESSURE:

$$\Delta p = \left(\frac{10\,780 - 10\,448}{10\,448} \right) \cdot 100 = 3.18 \ \%$$

DETERMINACIÓN DE LA PRESIÓN ESTÁTICA EN UNA PARTE DEL DIFUSOR CONDICIONADO UN DETERMINADO C_p:

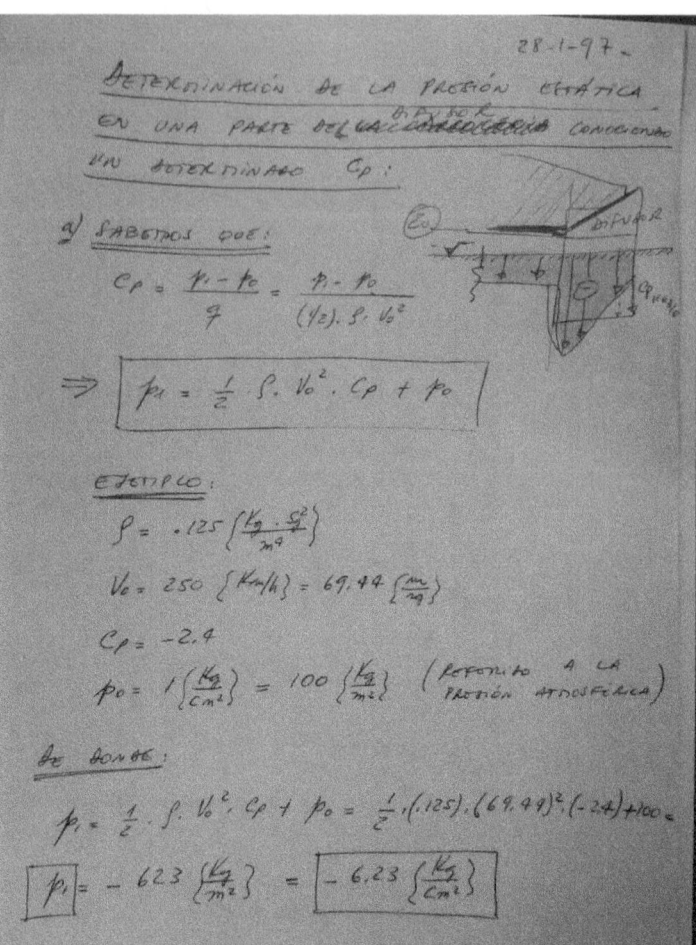

a) SABEMOS QUE:

$$C_p = \frac{p_1 - p_0}{q} = \frac{p_1 - p_0}{(1/2)\cdot \rho \cdot v_0^2}$$

$$\Rightarrow \boxed{p_1 = \frac{1}{2}\,\rho \cdot v_0^2 \cdot C_p + p_0}$$

EJEMPLO:

$$\rho = .125 \left\{\frac{kg \cdot \frac{s^2}{m^4}}{}\right\}$$

$$V_0 = 250 \left\{km/h\right\} = 69.44 \left\{\frac{m}{s}\right\}$$

$$C_p = -2.4$$

$$p_0 = 1 \left\{\frac{kg}{cm^2}\right\} = 100 \left\{\frac{kg}{m^2}\right\} \quad \left(\begin{array}{l}\text{REFERIDO A LA}\\ \text{PRESIÓN ATMOSFÉRICA}\end{array}\right)$$

DE DONDE:

$$p_1 = \frac{1}{2}\,\rho \cdot v_0^2 \cdot C_p + p_0 = \frac{1}{2}\cdot(.125)\cdot(69.44)^2\cdot(-2.4)+100 =$$

$$\boxed{p_1 = -623 \left\{\frac{kg}{m^2}\right\}} = \boxed{-6.23 \left\{\frac{kg}{cm^2}\right\}}$$

Now: special article:

High Efficiency Radiator Design for Advanced Coolant

Team 30
Brandon Fell – Recorder
Scott Janowiak – Team Leader
Alexander Kazanis – Sponsor Contact
Jeffrey Martinez – Treasurer

ME450
Fall 2007
Katsuo Kurabayashi

Abstract

The development of advanced nanofluids, which have better conduction and convection thermal properties, has presented a new opportunity to design a high energy efficient, light-weight automobile radiator. Current radiator designs are limited by the air side resistance requiring a large frontal area to meet cooling needs. This project will explore concepts of next-generation radiators that can adopt the high performance nanofluids. The goal of this project is to design an advanced concept for a radiator for use in automobiles. New concepts will be considered and a demonstration test rig will be built to demonstrate the chosen design.

Table of Contents

Introduction

Our task is to design an automotive radiator to work in conjunction with advanced nanofluids. The new radiator design will be used in new General Motors hybrid vehicles. These hybrid vehicles have multiple cooling systems for the internal combustion engine, electric engine, and batteries. The popularity of these hybrid vehicles is on the rise due to the decreasing fossil fuel supply, increasing the importance of a new radiator design that can possibly replace these multiple cooling systems.

Nanofluids

Nanofluids are a relatively new classification of fluids which consist of a base fluid with nano-sized particles (1-100 nm) suspended within them. These particles, generally a metal or metal oxide, increase conduction and convection coefficients, allowing for more heat transfer out of the coolant. There have been several advancements recently which have made the nanofluids more stable and ready for use in real world applications

Figure 1: TiO₂ Titanium Dioxide Nanofluid

These properties would be very beneficial to allow for an increased amount of heat to be removed from the engine. This is important because it will allow for a greater load to be placed on the fluid for cooling. However, these nanofluids do not show considerable improvement in heat transfer when used with current radiator designs. This is because there are several limitations to current radiator designs.

There are several basic requirements for this project. The new design must reject an increased amount of heat from current designs while lowering the inlet temperature. It must also have a more compact shape that will allow for alternate placement options within the vehicle.

This project is sponsored by Professor Albert Shih of the University of Michigan. We are in contact with Professor Shih's PhD student, Steve White [1], and will also collaborate with General Motors engineers further into the project.

Information Search

The first step of this project was to gather information on existing radiator designs and general heat exchangers. After gathering information, we gained a thorough understanding of how a radiator works and the disadvantages of the current radiator designs. This included a general patent search, using Google Scholar, and technical journal search, using Compendex, that related to radiators. Once we choose our design, we must research a general testing method to use as a basis for our comparison of our new design and the current designs.

Heat Exchangers

A steady-state heat exchanger consists of a fluid flowing through a pipe or system of pipes, where heat is transferred from one fluid to another. Heat exchangers are very common in everyday life and can be found almost anywhere [2]. Some common examples of heat exchangers are air conditioners, automobile radiators, and a hot water heater. A schematic of a simple heat exchanger is shown in Figure 2 below. Fluid flows through a system of pipes and takes heat from a hotter fluid and carries it away. Essentially it is exchanging heat from the hotter fluid to the cooler fluid.

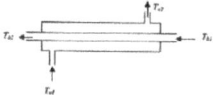

Figure 2: Simple heat exchanger

Automobile Radiators

Almost all automobiles in the market today have a type of heat exchanger called a radiator. The radiator is part of the cooling system of the engine as shown in Figure 3 below. As you can see in the figure, the radiator is just one of the many components of the complex cooling system.

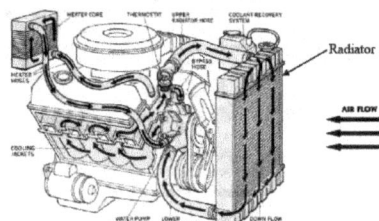

Figure 3: Coolant path and Components of an Automobile Engine Cooling System

5

Most commonly made out of aluminum, automobile radiators utilize a cross-flow heat exchanger design. The two working fluids are generally air and coolant (50-50 mix of water and ethylene glycol). As the air flows through the radiator, the heat is transferred from the coolant to the air. The purpose of the air is to remove heat from the coolant, which causes the coolant to exit the radiator at a lower temperature than it entered at. The benchmark for heat transfer of current radiators is 140 kW of heat at an inlet temperature of 95 °C. The basic radiator has a width of 0.5-0.6 m (20-23"), a height of 0.4-0.7 m (16-27"), and a depth of 0.025-0.038 m (1-1.5"). These dimensions vary depending on the make and model of the automobile.

For current radiator designs, a common configuration is to use parallel tubes which have aluminum fins attached to them. In these designs, there are basically three modes of heat transfer: conduction between tube walls and fins, and two modes of convection. One mode of convection is due to the coolant flowing in the tubes and the second is caused by the air flowing through the radiator. Associated with each type of heat transfer is a thermal resistance which obstructs the heat transfer rate. These resistances are summarized in Figure 4 below.

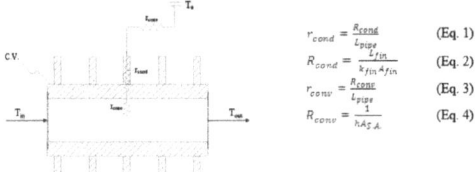

$$r_{cond} = \frac{R_{cond}}{L_{pipe}} \quad \text{(Eq. 1)}$$

$$R_{cond} = \frac{L_{fin}}{k_{fin} A_{fin}} \quad \text{(Eq. 2)}$$

$$r_{conv} = \frac{R_{conv}}{L_{pipe}} \quad \text{(Eq. 3)}$$

$$R_{conv} = \frac{1}{h A_{S,A}} \quad \text{(Eq. 4)}$$

Figure 4: A control volume thermal circuit diagram

Here, T_{in} represents the inlet fluid temperature, T_{out} represents the outlet fluid temperature, and T_a represents the ambient air temperature. As shown by Eq. 1, thermal resistance due to conduction per unit length (r_{cond}) is equal to the total resistance due to conduction (R_{cond}) divided by the length of the pipe (L_{pipe}). Eq. 2 provides the definition for R_{cond}. In this equation, L_{fin} is the length of the fin, k_{fin} is the thermal conductivity associated with the fin material, and A_{fin} is surface area associated with conduction. In this case, it would represent the bottom surface area of the fin. In Eq. 3, r_{conv} is equal to the total resistance due to convection (R_{conv}) divided by the length of the pipe. Here, R_{conv} is equal to 1 divided by product of the convective coefficient associated with the air (h) and the surface area exposed to the air ($A_{S,A}$). This can be seen by Eq. 4.

In current radiator designs, the largest thermal resistance is caused by the convective heat transfer (R_{conv}) that is associated with the air. This comprises of over 75% of the total thermal resistance. The second largest thermal resistance is caused by the convection that is associated with the fluid. Together, these resistances comprise of over 97% of the total thermal resistance [3]. Since there is a large thermal resistance associated with the air, the increased heat transfer

6

cannot be observed. Therefore, there is a need to design a radiator that reduces the percentage of thermal resistance associated with the air.

Limitations

Current radiator designs are extremely limited and have not experienced any major advancements in recent years. As described above, the main problem is that current radiators experience a large resistance to heat transfer caused by air flowing over the radiator. Current radiators also experiences head resistance, are very bulky, and impose limitations on the design of the vehicle.

Case Studies

After searching technical journals on Compendex, we found several related articles on different materials and designs for radiators. As shown in the case studies below, there are several ways to improve the current radiator design. This information will be used to develop a new design.

Case Study #1

Case study #1 showed that one way to decrease the thermal resistance associated with the air is to change the type of fin material used. Instead of using aluminum fins, fins constructed of carbon-foam were used. The fins were constructed out carbon-foam that had a porosity of 70%, a thickness of 0.762 mm, and a height of 8.725 mm. The fin density was set to 748 fins/m. The carbon-foam fins can be seen in Figure 5 below.

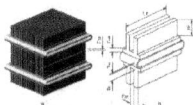

Figure 5: Carbon-foam Fin

Figure 6: Test setup for Carbon-foam Finned Radiator

The setup for this case study is shown in Figure 6 above. It showed that the percentage of thermal resistance associated with air-side convection was reduced to about 60%, therefore the percentage of the thermal resistance associated with the fluid was increased [3]. With the shift in these percentages, the convective benefits of nanofluids would have a more significant role.

7

Case Study #2

In case study #2, a possible improvement to the automobile radiator was seen through the analysis of micro heat exchangers. These heat exchangers incorporated the use of micro-channels and were fabricated from plastic, ceramic, or aluminum. The micro heat exchanger can be seen in Figures 7 and 8 below.

Figure 7: Micro-channel Heat Exchanger   **Figure 8: Micro-channel Heat Exchanger**

When compared to several automobile radiators, the micro heat exchanger outperformed them in a couple of areas. One area was on a heat transfer rate to volume basis in which the micro heat exchanger was better by more than 300%. Another area was a heat transfer rate per mass basis. In this area, the micro heat exchanger showed improvement of about 200%. These improvements were achieved by limiting the flow to smaller channels which increased the surface area/volume ratio and reduced the convective thermal resistance associated with the solid/fluid interface. However, in this study, the automobile radiators did outperform the micro heat exchanger on a heat transfer rate per frontal area basis. Here, the micro heat exchanger showed a reduction of over 45%. However, it is possible to construct a micro heat exchanger that has the same heat transfer rate/frontal area as current automobile radiators by using a more conductive material and reducing the spacing between the fins [4]. Therefore, when compared to automobile radiators, the use of micro heat exchangers allows the same amount of heat to be dissipated with a reduced volume and weight.

Case Study #3

In case study #3, the use of vortex generators was the technique used to improve the current radiator design. These incorporated wings on the fins which produced vortices that helped to increase the turbulence of the air. By increasing air turbulence, the convective coefficient associated with the air is increased. An increase in this value causes the thermal resistance associated with the air to be reduced. This can be seen by Eq. 4 on page 6. Figure 9 on page 9 shows the vortex generators in more detail.

Some parameters that affected the performance of the vortex generators were angle of attack, aspect ratio, and the ratio of vortex generator area to heat transfer area. With the use of the vortex generators, there was an increase in the convective heat transfer coefficient. Since the air-side resistance is directly related to this value (Eq. 4 on page 6), an increase in this value will decrease the thermal resistance due to the air [5]. Therefore, this configuration would be more beneficial than current radiator designs.

8

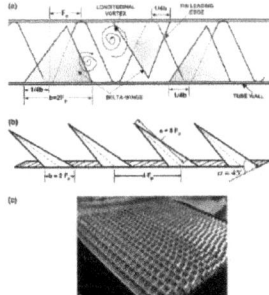

Figure 9: Vortex generators increase convective heat transfer

Customer Requirements / Engineering Specifications

After meeting with our sponsor, we have gathered the requirements for our project. Since this project is very open to radical designs, there are few quantitative requirements. We determined that dissipating a larger amount of heat to the air (147 kW), a smaller size (10-15% smaller in volume), lower inlet fluid temperature (to 85°C), and alternate placement options to be our customer requirements. However, the primary requirement is the increased dissipation of heat.

From the customer requirements, we determined the following items listed below in Table 1 to be our engineering specifications.

Engineering Specifications
- Dissipate 147 kW of heat total
- Decrease inlet fluid temperature to 85°C
- Decrease thermal resistance of air side by 5%
- Decrease total resistance of system by 5%
- Increase convective heat transfer coefficient of air by 5%
- Function with current hoses (1")
- Minimize frontal area 10-25%
- Minimize weight 10-20%
- Minimize flow rate 5-10%
- Minimize fluid capacity 15%

Table 1: Engineering specifications determined from customer requirements.

These specifications can be seen in conjunction with the customer requirements and benchmark products in the form of a QFD diagram in Appendix A on page 37. The specifications were

9

determined with help from benchmark numbers for heat transferred by the system and fluid inlet temperature for two different radiators given to us by our sponsor, along with a basic understanding of the heat transfer process for radiators. These represent our initial estimates as to the factors we believe will allow us to develop constraints for a theoretical model.

Concept Generation

In order to begin the design process, we began by breaking down the functions of the radiator. We did this in the Function Analysis System Technique (FAST) diagram shown in Figure 10 below.

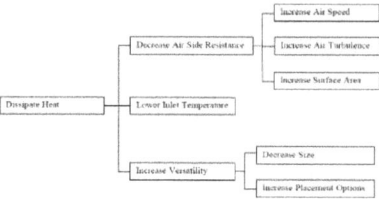

Figure 10: FAST Diagram

The basic function of the radiator is to dissipate heat from the coolant. This can be done by the subsidiary functions of decreasing air side resistance of the radiator and lowering the inlet temperature into the radiator. An additional subsidiary function is to increase the versatility of the radiator. Decreasing air side resistance can be done by the subsidiary functions of increasing the speed and lowering the temperature of the air flowing over the radiator and increasing the surface area of the radiator. Increasing the versatility of the radiator can be done by decreasing the size of the radiator, as well as change the shape.

We utilized the FAST diagram and researched background radiator information to brainstorm possible ideas and improvements that would help us achieve our primary requirement of dissipating 147 kW of heat. These ideas are displayed in the Morphological chart (Table 2) below on page 11.

New Radiator:	Rotary Radiator	Tube Cube [4]	Push-Pull Fans with Scoops	
Air Turbulence	Vortex Generator [3]	Dimples[1]	Offset Channels [3]	Carbon Foam[1,2]
Air Speed	Scoops[1]	Turbocharger [4]	Compressed Air	Airfoils [5]
Liquid Tubes	Smaller Tubes/Higher Density [4]	Increase Tube Width [1,5]	Dimples[3]	
Surface Area (Fins)	Carbon Foam [1,2]	Increased Thickness [4,5]		
Cool Inlet Fluid	Refrigeration Cycle [2]	Liquid Nitrogen		
Surface Area (Overall)	Cube Shape [4]	Wedge [5]	Swept Back Vertically [5]	

Key: 1 – Design 1; 2 – Design 2; 3 – Design 3; 4 - Design 4; 5 - Design 5

Table 2: Morphological Chart

Our ideas from the Morphological Chart fell into seven categories: new radiator designs, increase air turbulence, increase air speed, liquid tubes within the radiator, increase surface area of the fins, cool inlet fluids, and increase surface area of the radiator as a whole.

New Radiator Designs
We developed several new concepts for radiators. These ideas are full concepts addressing multiple issues. Our first idea in this category was a rotary radiator. This is simply a radiator rotating about a central axis. Coolant is pumped to the center, and through centripetal motion, is brought to the outside edge where it is collected and re-circulated to the engine. The fluid transfers the heat to the rotating structure through convection and since the structure is rotating at a high speed, the convection due to the air is increased. Our other idea for a new radiator was a tube cube. This design increases surface area. Tubes are bent and attached in a pseudo-random pattern that makes this design look like an aluminum tube cube. Our last idea for a new radiator design was push-pull fans with scoops. The concept was to increase the airspeed traveling through the radiator and to ensure the air passed all the way through it. Scoops would ensure air was being directed to the "push" fan which would send it through the radiator and get forced out by the "pull" fan.

Air Turbulence
We developed concepts that would increase air turbulence, thereby increasing convection. Our first idea in this category was vortex generators. Used in the aerospace industry, these small fins stick up from the airfoil surface at an angle to the direction of the airflow. Small vortices are created which keeps the flowing boundary layer of air on the wing surface longer through changing angles of attack. Applied to radiator fins, these "mini-tornados" increase the turbulence within the radiator, thus increasing heat transfer due to convection. Another idea was golf ball dimples. Along the same principal as the vortex generators, dimples have been applied to golf balls to keep the boundary layer attached to the surface longer, thus increasing drive distance. If applied to the radiator tubes, this would assist in increasing convective heat transfer. We also considered offset channels. By arranging the coolant tubes in an offset pattern, the air is forced to separate and weave around them. All this separation insures a large surface area available for convective heat transfer and a general disruption of the smooth airflow. Our last idea to increase air turbulence was the use of carbon foam. This relatively new material when

11

dried creates a virtual maze for the air to flow through. It also increases the surface area for both convective and conductive heat transfer.

Air Speed
In order to increase the speed of the air flowing over the radiator, we came up with several ideas. First, we considered adding scoops to the front of the radiator. These scoops would funnel air into the radiator and increase the velocity of the air by decreasing the cross sectional area of the scoop. This can be seen in the equation for fluid flow $Q = AV$, where Q is the volumetric flowrate of the fluid, A is the cross sectional area of the funnel, and V is the velocity of the fluid. While Q remains the same, A decreases, forcing V to increase. Our second idea to increase air speed was to use a turbocharger. A turbocharger is comprised of a turbine and compressor connected on the same axle. The inlet to the turbine is exhaust gases from the engine exhaust. This exhaust causes the turbine to rotate, which drives the compressor. This compressor then blows out air at a high velocity. We had another idea involving the turbocharger. Instead of having the turbocharger blow air onto the radiator directly, the turbocharger would compress air into a pressure vessel. This pressure vessel would hold the compressed air and release it onto the radiator in timed bursts. Our last idea in this category was airfoils. These airfoils increase the velocity of air.

Radiator Tubes
Our next category was the tubes in the radiator that carry the coolant. Our first idea in this category was to make the tubes smaller and increase the total amount of tubes. This concept decreases the time it takes to transfer the same amount of heat by exposing it in more places within the radiator. Another idea was to increase the width of the tubes. By increasing the width of the tubes, we increase the surface area of the tube. This increased surface area allows for more heat transfer by convection. Our last idea was dimples. These dimples create air turbulence similar to that of golf balls.

Surface Area (Fins)
In order to increase the surface area of the fins, we came up with two ideas. First, we came up with the idea to increase the thickness of the fins. The surface area associated with the fins includes the top, bottom, front, and back. Therefore by making the fins thicker, the surface area is increased by increasing the front and back areas. We also decided that the use of carbon foam mentioned above increases the surface area of the fins. This is because the carbon foam is porous and allows the air to flow through it in addition to flowing around it.

Inlet Fluid
Another category on our morphological chart was to cool the inlet fluid. Our first design concept in this category was to use the refrigeration cycle. This would utilize a fluid – fluid heat removal system, which would pull more heat from the engine than a liquid-air system. By removing more heat from the coolant, the inlet fluid temperature was reduced. Another design idea to lower fluid inlet temperature is to atomize liquid nitrogen. This would create a larger temperature difference between the nitrogen and the coolant because the liquid nitrogen is at a much cooler temperature than the ambient air. With this increased temperature difference, more heat could be removed from the coolant.

Surface Area (Total)

Our last design category was to increase the overall surface area of the radiator. The first design idea in this category was to make the radiator a cube shape. A cube is compact and has a large surface area-to-volume ratio. Another design idea to increase overall surface area was to make the front of the radiator a wedge. The projected area of the two sides would increase compared to current radiator designs. Our last design idea in this section was to sweep the radiator back vertically. The overall surface area is increased by adding more depth to the radiator design.

Concept Evaluation and Selection

After reviewing some of our design ideas, we used our morphological chart to combine multiple design ideas into five design concepts. We attempted to include design ideas from different categories and ensured the feasibility of combining the various design ideas. Five of our best design concepts are listed below.

Design Concept #1

Concept sketch #1, shown in Figure 11 below, incorporates the use of golf ball type dimples on the surface of the coolant tubes. By creating a rough surface, these dimples aid in increasing the air turbulence. By increasing the air turbulence, the convection coefficient increases. Therefore, the resistance due to the air-side convection is reduced as seen in Eq. 4 on page 6. In addition to the dimples, this design also incorporates the use of air scoops that channel the incoming air into the radiator. This aids in increasing the velocity of the air. With the increased air velocity, the convection coefficient associated for the air is increased. By increasing this value, the thermal resistance associated with the air is decreased. This can also be seen in Eq. 4 on page 6. Also included in this design is increased tube width and fin thickness. By increasing these dimensions, more surface area is exposed to the incoming air. Due to the increased exposure, the thermal resistance associated with the air is decreased. This can also be seen in Eq. 4 on page 6. This design also used carbon foam fins which replaced the aluminum fins used on current radiator designs. The carbon foam also increases the surface area exposed to the air. This is mainly due to the fact that the carbon foam is porous and allows the air to flow thru it in addition to allowing the air to flow around it.

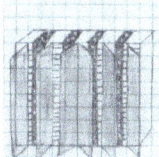

Figure 11: Concept #1

The benefit of this design is that all of the changes to the current radiator design help to reduce the thermal resistance associated with the air. This is either done by increasing the surface area

13

exposed to the air or increasing the convection coefficient. There are several drawbacks associated with this design. One drawback is that the design still relies heavily on the air. Therefore, there is always going to be a thermal resistance associated with the air. Another drawback is that the increased material used for the fins, tubes, and scoops will increase the cost of the radiator. In addition to this, carbon foam is more expensive than aluminum and will also increase the cost of the radiator. Also, since the carbon foam is porous, it is susceptible to becoming clogged by bugs and other environmental debris. Therefore, the carbon foam would require periodic cleaning by the owner in order to maintain the benefits associated with this material.

Design Concept #2
Concept sketch #2, shown in Figure 12 below, illustrates the refrigeration cycle, which we would use to replace the radiator. This would be an additional refrigeration cycle from the cycle already existing in the vehicle. This would eliminate the dependence on air to cool the coolant. This cycle incorporates the use of a dual fluid heat exchanger. The purpose of the heat exchanger is to remove heat from the engine coolant by adding it to the refrigerant, R-134a. Once this is done, the refrigerant gets compressed in the compressor and then moves on to the condenser. In the condenser, the refrigerant loses the heat it received from the engine coolant. Then, it passes through the expansion valve and then through the evaporator. Once it passes through the evaporator, it enters the heat exchanger and the cycle repeats itself.

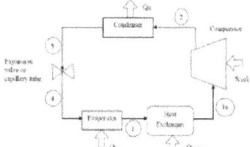

Figure 12: Basic Refrigeration Cycle with Additional Heat Exchanger

One benefit of this design is that it reduces the dependence on the air to cool the engine coolant by using the refrigerant. By using the refrigerant, we would be able to remove more heat from the coolant than by using a liquid-air heat exchanger. One drawback with this system is the added cost associated with the refrigeration system components. Another drawback is the complication of placing the various components of this system within the engine compartment.

Design Concept #3
Concept sketch #3, shown in Figure 13 below on page 15, is a radiator with tubes in the shape of an airfoil, or wing. The radiator would have tubes in the shape of a wing. This would increase the velocity of the air over the radiator itself. This is because the air increases velocity travelling over the airfoil in combination with the pressure drop over it. Because the top of the airfoil has a larger surface area, the velocity over the top would have to increase in order to meet the air

14

flowing over the bottom at the same time. The front of the design is wedge-shaped. The shape of the wedge increases the overall surface area of the front of the radiator. This increases the convection coefficient and allows for more heat transfer to occur out of the radiator. Vortex generators were also added to create turbulence. This also increases the convection coefficient, allowing for more heat transfer out of the radiator. Similarly, this design also has golf ball dimples. This will also increase the convection coefficient.

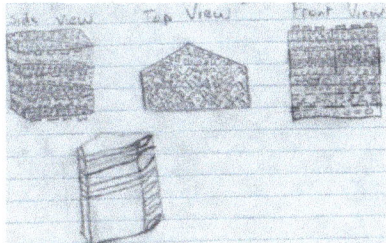

Figure 13: Wedge concept

While this design increases the radiator's convection coefficient, the bulky shape hinders alternate placement options. In order to maintain the same volume as current radiators, the volume cut off in the front to create the wedge would have to be added to the back. This would increase the overall depth of this design as compared to a current radiator. While the front of the radiator has a greater surface area, because there is space between the tubes themselves, overall surface area decreases in this design. This decrease in overall surface area negates the increase in the convection coefficient, and results in less heat transfer and lower the fluid inlet temperature compared to standard radiators.

Design Concept #4
Concept sketch #4, shown in Figure 14 on page 16 below, illustrates our tube-cube idea. The motivation for this concept was to maximize the surface area of the coolant tubes in a cubic shape. We chose the cubic shape because it allows for the maximum volume with the smallest side length dimensions. We also chose to employ a variable geometry turbocharger to increase the air flow across the tubes. We chose a variable geometry turbocharger to provide a constant air flow during idle and low engine speeds. As seen in the figure, the tubes run parallel to each other in a plane. They are then off-set in the next plane to create turbulence in the air flowing across the tubes. The variable turbocharger blows air into the center of the cube in order to carry the heat out of the cube on all sides.

15

Figure 14: Turbo tube cube concept

The large tube surface area combined with increased air flow allows for a reduced convective resistance for air-side cooling. This concept also provides several alternative placement options for an automotive application. This concept will replace the radiator fan found on current vehicles, with a variable geometry turbocharger. However there are a few drawbacks to this concept. There will be an increased amount of pumping work required from the water pump in order to pump the fluid through the many bends in the cube. Also, the variable geometry turbocharger will add cost to the total cost of the radiator.

Design Concept #5
Concept sketch #5, shown in Figure 15 below, illustrates our concept of a stretched-back radiator. The concept for this design was to maximize the surface area that the air came into contact with. Recent developments in the application of carbon foams motivated the use of them in this idea. From the Pugh chart above, offset channels, carbon foam, increased thickness (depth), and a vertical sweep back were combined.

Figure 15: Carbon-Foam channel concept

Having offset channels helped to increase the number of passes the fluid would have to make through the radiator thereby increasing the temperature difference between the coolant inlet and outlet points. They also assisted in creating a compact design, one of the requirements set by the customer. The carbon foam would have been cast into the desired shape and would structurally support the channels for the coolant. The desired shape was swept back to increase the surface area available to the air to maximize convective heat transfer.

While this concept holds the possibility of reaching the cooling requirements set forth by our customer, it would be rendered useless in an automotive application in a matter of days. Due to the small pore size in the carbon foam, foreign particles, insects, and other items would

16

conglomerate within the radiator and decrease effectiveness to the point that it would cease to function.

Selected Concepts

We used a Pugh Chart (Table 3) below to evaluate our different design concepts. We weighted the importance of the customer requirements, then evaluated whether the design concepts outperformed the current radiator models, in which case it was given a plus. If the concept performed worse than the current model, it received a minus. The concept received a 0 if there was no difference. Cost was not an initial consideration because our budget was $800, and the customer was focused more on increasing heat dissipation and reducing size. After our evaluation process, we narrowed our final design down to the two concepts listed below.

Customer Requirements	Weight	Sketch 1	Sketch 2	Sketch 3	Sketch 4	Sketch 5
Dissipate more heat to the air	10	+	+	-	+	+
Smaller size	6	-	+	-	0	-
Lower fluid inlet temperature	3	+	+	-	+	+
Alternative placement options	1	-	0	-	+	+
	$\sum +$	13	19	0	14	14
	$\sum -$	7	0	-20	0	6
	$\sum$ total	6	19	-20	14	8

Table 3: Pugh Chart

Sketch #2 – Refrigeration Concept

The cycle for our refrigeration concept can be seen below in Figure 16 on page 18. This cycle varies slightly from the conventional refrigeration cycle because we have replaced the evaporator with a heat exchanger to pull the heat from the coolant of the radiator. This heat exchanger will still do the job of the evaporator by heating the working fluid to a gas. We will also still be using a conventional radiator, but we can reduce the size of this radiator. The purpose of leaving a radiator in the vehicle is to reduce the amount of compressor work required to properly remove heat from the engine and heat rejected by the condenser. This would reduce the dependency of the radiator on the air side cooling.

For this design, we plan to use ¼" diameter tubes in order to achieve our desired estimated flowrate of 2.5 kg/s. The flowrate of 2.5 kg/s was chosen to allow us to achieve the desired amount of heat transfer. Also, 2.5 kg/s is feasible to apply to our system. We would like to use an Exergy, LLC miniature heat exchanger.

17

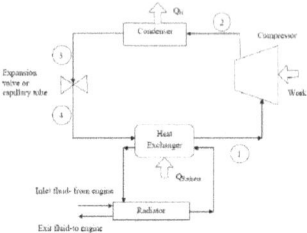

Figure 16: A/C Radiator concept

Sketch #4 – Turbo Tube Cube Concept

The turbo tube cube concept allows for a large surface area to dissipate heat combined with a turbo-charger to increase airflow over the coolant tubes. These two design ideas help to lower the air-side resistance. The tubes will be made out of aluminum or aluminum alloys to keep material and manufacturing costs low. The turbocharger will have variable geometry in order to keep consistent airflow during idle and low engine speeds. In order to get a grasp on how big this concept would really need to be, we used a simple mathematical model to estimate the total surface area needed to reject 147kW of heat.

For the simple mathematical model, we assumed a straight tube with a constant rate of cooling. We used Newton's Law of Cooling (Eq. 5 below) to estimate the surface area required to dissipate the 147kW of heat. In this equation Q = 147kW, h is the convection of the fluid (we assumed h=100 W/m^{2}K), and T_0 is the fluid temperature and T_a is the ambient temperature (assumed to be 25°C). We assumed h because that is the value we are expecting for our model (we are unable to calculate it explicitly because we do not have the dimensions of the design).

$$Q = hA(T_0 - T_a) \qquad \text{(Eq. 5)}$$

After running through the calculations, we found a simple cube with round tubes to be too large to meet our requirements. Therefore we changed our design slightly to compensate for the required surface area. We employed a finned tube design which gave us two to three times the surface area on the tubes. We also diverted from the cubic design to a more rectangular design with bent tubes that allowed us to make the tubes slightly longer without increasing the overall length of the radiator. The bent tubes also allow us to focus the turbocharger into the inner curved section and achieve fairly consistent airflow over the tubes. Figure 17 on page 19 shows the modified tube cube design. A dimensioned drawing can be found in Appendix C on page 39, and front, right, and top drawings can be found in Appendix D on pages 40 and 41.

18

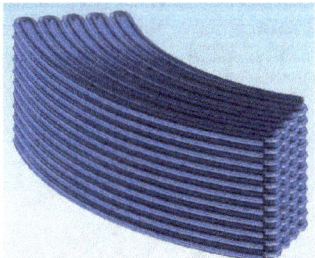

Figure 17: Isometric CAD model of the Tube Cube Concept.

Design Evolution

Upon further consideration, both of our selected concepts were infeasible. The tube cube concept required too many passes of the coolant tubes, making the new design much larger than current designs. Also the increase in pumping work would be required. The refrigeration concept would require a larger condenser to achieve the required heat rejection. The work input from the compressor would also be increased, which would cause a parasitic loss of the engine.

Reconsideration
Due to the infeasibility of our selected concepts, we decided to reconsider the use of carbon foam in our design. Carbon foam is a porous foam, which is made from coal. When heated in excess of 2000°C, the carbon takes the form of graphite, which is the primary material in carbon (graphite) foam.

Advantages
Carbon foam provides a large surface area per unit volume due to large and numerous pores. This large surface area will increase the surface area exposed to the air and thus reduce the air side resistance. Carbon foam is very lightweight when compared to conventional materials used in current radiators (aluminum or copper). It can also be manufactured from a block to any desirable shape by means of milling, cutting, drilling, etc. Carbon foam also is a sponge-like material, which is more durable compared to aluminum fins.

Disadvantages
The major disadvantage associated with carbon foam is that it is expensive to produce, with a commercial cost around $5.00 per cubic inch. However, new production methods show potential to lower the price in the near future. Also, the many small pores in carbon foam can become clogged with road debris or insects, but a filtering screen should keep the foam clean for our application. It also requires additional bracing for support.

New Design Concept

Our new design concept is similar to current radiators, but replaces aluminum fins with carbon foam channels. Due to the thermal properties of carbon foam (k = 175-180 W/mK for carbon foam with 70% porosity), along with increasing the amount of heat rejected, we will be able to reduce the overall size of the radiator while simultaneously increasing the surface area exposed to the air, thus reducing the air side resistance. Figure 18 below shows our new design concept.

Figure 18: Carbon Foam Radiator Concept

The carbon foam has channels in a corrugated pattern. This corrugation channels air into the slots and forces the air through the carbon foam. Also, there are many tubes which are arranged in a parallel design. They provide support for the carbon foam as well as contain the necessary volume of coolant. The end caps are made out of aluminum and also provide structural support and mounting locations. Overall, this design concept is a simple design which will meet most of our customer requirements, including dissipating 147 kW of heat with an inlet fluid temperature of 85°C, decreasing the overall volume.

Engineering Analysis

Initial Calculations

A preliminary CAD model was constructed with a height of 10", a length of 15', and a depth of 1.5" as shown in Figure 18 above.

A cross-sectional diagram of a radiator section displaying the tube configuration can be seen in Figure 19 below on page 21. The five tube array was repeated 19 times, resulting in 95 tubes being used for the preliminary concept model.

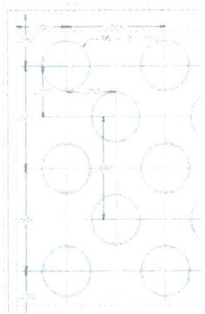

Figure 19: Cross Section of C.F. Radiator Concept

In Figure 20 below on pages 21 and 22, a schematic for the carbon foam section used for analysis can be seen. Figure 20 (a) shows a sample section on the preliminary CAD model. Figure 20 (b) shows the isolated sample section, displaying only one tube because the repeated array of tubes (average of 2.5 tubes per row) was lumped together to give one tube with a diameter of 0.625" (2.5 x ¼"). This assumption underestimates the total heat transfer, due to the fact that the overall surface area exposed to the air is reduced. Therefore, we expect our test results for heat transfer to exceed those calculated in this analysis. This section model is 0.5" long, repeating 30 times, summing to the overall length dimension of 15". The height of the section model was 0.526", repeating 19 times, summing to the overall height dimension of 10". The depth is the same as the preliminary CAD model.

(a)

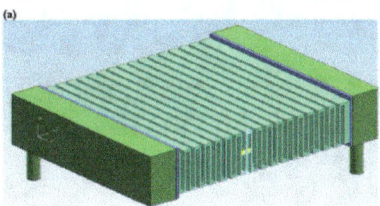

(b)

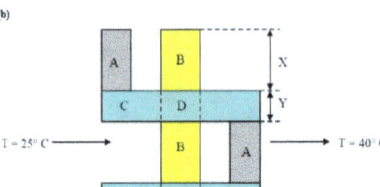

Figure 20: Carbon foam section used in analysis (a) as seen on CAD model and (b) model showing isolated section

The preliminary model was used as a starting point for the final design. We used a thermal circuit analogy for our system because it was modeled as one dimensional and was under steady-state conditions. The thermal circuit analogy was used to determine dimensions X and Y seen in Figure 20 (b) above. The thickness of the carbon foam sections is represented by Y and the length of the bare tube exposed to the air is represented by X. In our model, we assumed the tubes to be thin walled. Therefore, we neglected the thermal resistance due to conduction through the tube wall. It is also important to note that we assumed that the air flows through the section; however, due to the corrugated pattern, additional air flow would result because the air is forced into adjacent channels.

Next, we determined the thermal resistances for each part of the thermal circuit. Figure 20 above also shows the letters corresponding to the sections that will be discussed in the following paragraphs.

Section A

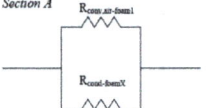

In this section, there are two resistances: $R_{conv,air-foam 1}$ and $R_{cond-foam x}$. These resistances are associated with convection and conduction through the carbon foam, respectively. In order to determine $R_{conv,air-foam 1}$, the average convection coefficient associated with A had to be determined. This section was modeled as a vertical plate with a height of 0.00085 m. Using Eq. 6 below, the Reynolds number was 1348.

$$Re_D = \frac{\rho V D}{\mu} \qquad \text{(Eq. 6)}$$

22

In this equation, ρ represents the air density, V represents the air velocity, D represents the cross sectional height, and μ represents the dynamic viscosity of the air. In Eq. 7 below, k represents the thermal conductivity of the air and Pr is the Prandtl number. The values for the constants C and m were 0.228 and 0.731, respectively [6]. All of the air properties in Eq. 6 and 7 were evaluated at the film temperature of 333 K.

$$\overline{Nu_D} = \frac{\bar{h}}{k} D = C \operatorname{Re}_D^m Pr^{1/3} \qquad \text{(Eq. 7)}$$

The average convection coefficient, $\bar{h}$, was 391 W/(m^{2}K). However, typical values of $\bar{h}$ for forced convection using gases typically range from 25-250 W/(m^{2}K). Therefore a value of 150 W/(m^{2}K) was estimated by comparing this configuration to configuration B and deciding that $\bar{h}$ should be lower for this section. This was because the air flow around the tube would remove more heat than the air flow over the foam. The airflow around the tube completely encompasses the tube, whereas the flow over the foam would tend to separate.

Once $\bar{h}$ was established, $R_{conv,air-foam 1}$ could be determined. This value was simply $1/(\bar{h} A)$, where A is the surface area exposed to the air.

$R_{cond-foam 1}$ was determined by using the equation for the thermal resistance of a slab, $L/(kA)$. Here, L is the thickness of the foam, k is the thermal conductivity of the foam, and A is the frontal area, which is $1/3" \cdot X$.

Section B

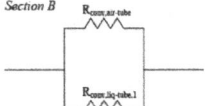

This section also has two resistances: $R_{conv,air-tube}$, due to the air flowing through the tube, and $R_{conv,liq-tube 1}$, due to the coolant (water) flowing through the tube. In order to determine both of these resistances, a convection coefficient, h, had to be determined for each type of convection. To determine h for the air, the tube was modeled as a cylinder in cross flow. In this case, D was the diameter of the tube and all of the values of the air properties, as well as the velocity, remained the same as described in the analysis of section A. Also, the values for C and m were 0.683 and 0.466 respectively [6]. Evaluating these equations resulted in an $\bar{h}$ of 180 W/(m^{2}K). Using this value, we were able to obtain $R_{conv,air-tube}$ by using the formula $1/(\bar{h} A)$, where A is the surface area of the tube exposed to the air ($D\pi X$).

To determine h for $R_{conv,liq-tube}$, Eq. 8 and 9 were used. In Eq. 8 on page 24, (m) is the mass flow rate which was 0.063 kg/s. This was determined by using a flow rate for a typical radiator. The inlet flow rate for a typical radiator (while the vehicle is traveling at 29 m/s) is approximately 2.4 kg/s. In our initial design, we had 95 tubes. Therefore, the flow rate for one tube in the preliminary CAD model was (2.4 kg/s)/95, because the tubes were in parallel. For our model tube (2.5 tubes), we multiplied this value by 2.5. D represents the diameter of the tube and μ is

23

the dynamic viscosity of the water. In Eq. 9, k is the thermal conductivity of the water and n is equal to 0.3. This is because the surface of the tube is cooler than the mean fluid temperature of the fluid (344K). In both equations, all fluid properties were evaluated at the mean fluid temperature. Eq. 9 is valid because the total length of the tube divided by the diameter is greater than 10, which causes the flow to be fully developed [6]. In addition, Re_D was greater than 10,000 which meant the flow was turbulent [7]. Also, Pr was in between 0.6 and 160.

$$Re_D = \frac{4\dot{m}}{\pi D \mu} \qquad \text{(Eq. 8)}$$

$$Nu_D = \frac{hD}{k} = 0.023\, Re_D^{4/5}\, Pr^n \qquad \text{(Eq. 9) [6]}$$

The value for h obtained in Eq. 9 was 2466 W/(m²×K). Using this value, we were able to obtain $R_{conv,liq-tube}$, equal to $1/(hA)$, where A is πDX.

Section C

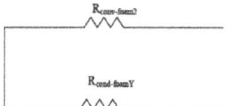

This section has two associated resistances: $R_{conv,foam\,2}$ and $R_{cond-foam,\,y}$. $R_{conv,foam\,2}$ is the resistance to convection over the foam. This resistance is also associated with section D, described below. This is because the convection over the foam occurs simultaneously over sections C and D. $R_{cond-foam,\,y}$ is the resistance to conduction through the foam. In order to determine $R_{conv,foam\,2}$, $\bar{h}$ had to be established. To accomplish this, the section was modeled as a flat plate in external flow and Eq. 10 and 11 were used.

$$Re_L = \frac{VL\rho}{\mu} \qquad \text{(Eq. 10)}$$

$$\overline{Nu_L} = \frac{\bar{h}L}{k} = 0.680\, Re_L^{1/2}\, Pr^{1/3} \qquad \text{(Eq. 11)}$$

In Eq. 10, L is the overall length of the foam, from the front face to the back, V represents the air velocity, ρ represents the air density, and μ represents the dynamic viscosity of the air. In Eq. 11, k is the thermal conductivity of the air and Pr is the Prandtl number. In both equations, all air properties were evaluated at the film temperature (333K). Eq. 10 established that the flow was laminar because Re_L was less than 5x^5. Therefore, Eq. 11 was used. In using this equation, we also assumed that the surface provided a uniform surface heat flux rather than a uniform temperature [6]. The value of $\bar{h}$ is 110 W/(m²×K). The associated thermal resistance is $1/(\bar{h}A)$, where A is the surface area exposed to the air.

The expression for the second resistance, $R_{cond-foam,\,y}$, is $L/(kA)$. Here, L is the length of the foam, k is the thermal conductivity of the foam, and A is the cross sectional area of the carbon foam. In calculating this resistance, the carbon foam was modeled as a solid piece with the tube section at D removed.

Section D

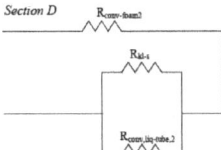

There are three thermal resistances associated with section D. They are $R_{conv,foam\,2}$, $R_{k,ls}$, and $R_{conv,liq-tube\,2}$. These are due to the convection caused by the air flowing over the foam, the radial conduction from the outside of the tube to the surface of the carbon foam, and the convection due to the water flowing in the tube respectively. The resistance $R_{k,ls}$ is defined by Eq. 12 below.

$$R_{k,ls} = \frac{\ln(r_2 / r_1)}{2\pi L k} \qquad \text{(Eq. 12) [6]}$$

In using this equation, the carbon foam around the tube was modeled as a cylinder instead of the rectangular shape it actually was. This allowed us to continue the use of the one dimensional model. In this equation, r_2 and r_1 are the radii of the outside surface of the carbon foam and the outside surface of the tube, respectively. L is the length of the tube section, which we defined as Y, and k is the thermal conductivity of the carbon foam.

The analysis for $R_{conv,liq-tube\,2}$ is the same as $R_{conv,liq-tube\,1}$ except that the value for A changes. Instead of πDX, it is πDY.

Upon inspection, we determined that the resistances for section A and B were then repeated, except B now came before A. Then, the resistances associated with sections C and D were simply repeated, being the exact same as the analysis above. Having formulated all of our resistances for the section model, we combined them into the circuit diagram for the section model shown in Figure 21 below on page 26.

25

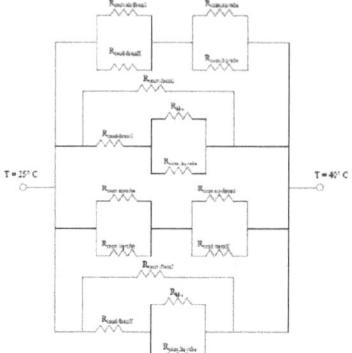

T = 25° C   T = 40° C

Figure 21: Thermal circuit used in analysis

The thermal circuit diagram was then simplified to obtain one resistance, R_{tot}. This was done by combining resistances in parallel and in series by using Eq. 13 and 14 respectively. These equations were simplified utilizing Maple™. A summary of the calculated values can be found in Appendix E on page 42, and a printout of the Maple™ code can be seen in Appendix F on pages 43 through 47.

$$\frac{1}{R_{eq}} = \frac{1}{R_1} + \frac{1}{R_2} \qquad \text{(Eq. 13)}$$

$$R_{eq} = R_1 + R_2 \qquad \text{(Eq. 14)}$$

$$R_{tot} = \frac{\Delta T}{q} \qquad \text{(Eq. 15)}$$

Once this was done, Eq. 15 was used [6]. Here, ΔT is the difference in the air temperature of the air exiting the radiator and entering the radiator. We assumed a value of 40°C and 25°C for an exit and entrance temperature respectively. q is the desired heat transfer rate of 147 kW given to us by our sponsor. However, the required heat transfer rate for the modeled section was 147kW/1140. The heat transfer was divided by 1140 because the section modeled is repeated 30 times to get the length dimension and 38 times to get the modeled number of tubes (95). Using Eq. 15, our target value for R_{tot} was found to be 0.116 K/W. R_{tot} found above, as a function of X and Y, was equated to 0.116 K/W. We set X equal to 0.00254 m and Y was calculated by trial and

26

error. The value for Y' was 0.00205 m. However, we feel that this dimension is an underestimate due to the fact that there was no way to calculate the surface area of the pores in the carbon foam. Using the larger surface area would in turn lower the resistance of the system, resulting in a larger length of carbon foam. The overall dimensions of the radiator can be seen in Figure 22 below on page 28.

This analysis does not take into account the pressure drop of the airflow through the foam. This pressure drop is difficult to model due to the varying pore sizes. We did not model this because we assumed it would be small and negligible in our design. We did not have the proper knowledge to model it and time constraints forced us to simplify the model. In reality, the pressure drop would have an effect on the airflow and the heat transfer.

Corrected Calculations
Due to the unexpected performance of the prototype, a reevaluation of the engineering analysis was conducted. Several numerical errors were found in this section. The first error was found in section A. Instead of a vertical plate with a height of 0.00085m, it should have been a vertical plate with a height of 0.013m. This is the result of the total height of 10" divided by 19 sections. This would result in a Reynolds number of 21000. Using this value, the average convection coefficient would be 620 W/(m^2K). In addition to this, the area used would be was also changed due to the height of the plate. The thermal resistances can be seen in the Appendix E on page 42. Also for this section, there was an error in determining $R_{cond-foam\,x}$. In this analysis, a conduction coefficient of 1200 W/mK was previously used. After gathering additional information, it was realized that this value should have been approximately 175 W/mK. Also, the area used in this calculation would change since the height of the plate changed.

In section C, there is an error associated with $R_{conv-foam\,2}$. It is due to the area that was used. In the previous analysis, the height was 0.00846m. However, it should have been 0.013m. This change in height would in turn cause the area used to differ from the previous value. The change in area would result in a different value of $R_{conv-foam\,2}$.

In addition to this, the value for $R_{cond-foam,y}$ was also in error. This resulted from using the wrong value of the thermal conductivity in the previous analysis. Also, as mentioned previously, the wrong height was used. This resulted in the incorrect area being used which in turn resulted in obtaining the incorrect value for $R_{cond-foam\,y}$.

In section D, there was an error associated with R_{al-s}. This was also due to the wrong height being used. In this case, the height corresponded to the outer radius (r_2). In addition to this, the wrong value of the thermal conductivity was previously used.

After using the corrected analysis, it was found that the value for Y, the thickness of the foam, should have been 0.0005 m (0.02''). This was obtained while keeping X at 0.00254 m. This would result in a radiator with a length of 0.0912 m (3.59''), a height of 0.0254 m (10"), and a depth of .00375 m (1.5").

Materials & Tolerances

Aluminum was picked as the material of choice because it is cheap, easy to work with, and the traditional material used today in radiator manufacturing. We chose to use 6061-T6 grade aluminum because of its versatility. It is easy to form, has good corrosion resistance, can be welded, and has medium strength.

The carbon foam was chosen for the radiator "fin" material because of the increased surface area available to cooling. Due to the material being relatively new, it is expensive and requires a complicated manufacturing procedure. Koppers was chosen as the supplier of the foam because of their recent advances in the manufacturing process of the foam.

All tolerances in the manufacturing process will be ±0.005". We chose this tolerance because it allows for accuracy and variations in the various processes used.

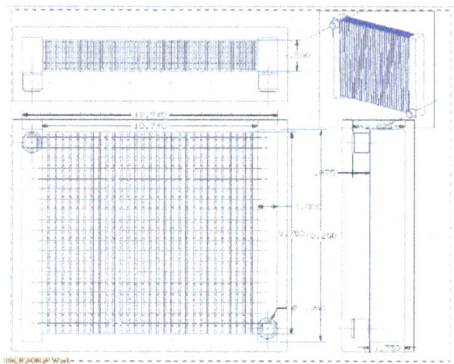

Figure 22: Final Dimensions of Carbon Foam Radiator

Final Design

After doing a complete engineering analysis, we now have our final dimensions. The overall dimensions of the carbon foam are 10.74" x 9.75" x 1.5", as found in Figure 22 above. This figure also shows the overall dimensions of the radiator including the end manifolds. These dimensions are 12.74"x 10.25" x 1.75". The final design also uses 1/4" aluminum tubing and 1/8" aluminum plating welded together to form the fluid path through the carbon foam. The inlet and outlet tubes are 0.95" diameter. The tubes are all parallel to each other and are mounted

28

horizontally. They also support the carbon foam and provide mounting locations for placement in the vehicle.

The carbon foam follows the corrugation design as mentioned earlier. The corrugation along with detailed dimensions can be found in Figure 23. These dimensions were determined through the engineering analysis described in the previous section.

The model, as designed, has an array of 95 - ¼''-diameters, 11.24'' long aluminum tubes to maximize the volume of fluid exposed to the air and carbon foam per unit time. The tubes are arranged in a staggered form as shown in Figure 24 below. Appendix G on pages 48 and 49 contains the complete set of CAD drawings of the radiator.

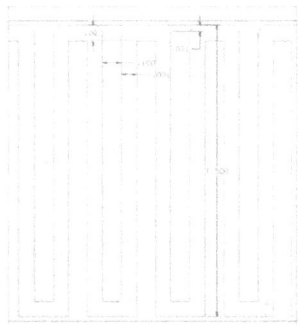

Figure 23: Corrugation Dimensions of Carbon Foam Radiator

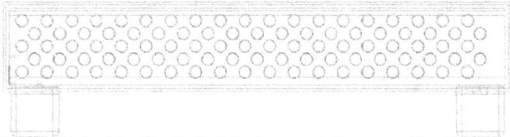

Figure 24: Side view of the radiator showing the 95 tube array maximizing exposure of the coolant.

29

Our sponsor requested that we build a section of the full model for proof of concept (POC) validation. To this end we have an $800 budget to purchase the necessary materials. A summary of the materials that are required for the POC can be found in Table 4, including the price and quantity of aluminum needed for our model. Special thanks to Thomas Golubic of Koppers for providing several blocks of KFOAM™ for use in this project. Once the POC is built, we will extrapolate the total amount of heat rejected from the test data, and compare the results to our original model.

Bill of Materials				
Quantity	Part Description	Purchased From	Part Number	Price (each)
	Carbon Foam	Koppers Inc.		$0.00
6	1.4"OD x 72" 6061 T6 Aluminum Tubing	Onlinemetals.com		$11.43
1	3.8"OD x 12" 6061 T6 Aluminum Tubing	Onlinemetals.com		$2.85
1	12" x 24" 6061 T6 Aluminum Sheet	Onlinemetals.com		$11.42
1	Turbine Flow Meter	Instrumart.com	G2A05N09LMA	$260.00
	Miscellaneous	Stadium Hardware		$40.00
2	Thermocouples	4oakton.com	WD-08500-55	$40.00
			Total =	$462.85

Table 4: Bill of Materials

Our design meets most of the engineering requirements set by our sponsor. The primary goal of improving heat transfer by 5% (147 kW) should be met according to our calculations. Also the inlet fluid temperature of 85°C should be met. According to our calculations, we were able to reduce the overall volume over 50%, which is much greater than our initial goal of a 10-25% reduction. By using carbon foam, we increase the surface area exposed to air, and therefore reduce the thermal resistance of air. We were also able to decrease the frontal area over 50%, which is much greater than the initial goal of 10-25%. By using carbon foam, we also reduce the weight of the radiator by more than 25%. Also, our design is able to function in its current environment.

Some of the issues that need to be addressed in future work are the cost and clogging of the carbon foam. Due to the recent developments in the production methods of carbon foam, we feel that the cost will be reduced in the near future. In order to keep the carbon foam from becoming clogged with road debris and insects, we propose a filtering screen be placed in front of the assembly. This solution would require the user to clean the screen at regular intervals.

Manufacturing Plan

Proof of Concept (POC)
To demonstrate POC for the theoretical model developed above, a 6" x 6" x 1.5" section will be built. This POC will represent approximately 25% of the theoretical model, and, assuming a linear relationship of area to heat transferred, it will be assumed to transfer 33% of the 147 kW sought, or about 48 kW.

30

Ideal process

Prior to any machining operations with the carbon foam, the end manifolds of the radiator must be formed and each aluminum tube cut to size. One side of each manifold will have a negative of the tube array in it so that once the tubes are inserted into the foam; the manifold can be welded to each tube ensuring a leak-proof enclosure. The manifold itself will be constructed from sheet aluminum bent in a break and welded at the seams. Hose fittings will be welded at appropriate points to allow fluid in and out of the radiator. The carbon foam, while able to be machined, will undergo the least amount of machining. The tubes can be inserted by hand or through a press using a stencil or the back-plate of either manifold into the foam and then removed. Once the tube array has been created, Wire Electrical Discharge Machining (EDM) will be used to create the air channels in this corrugated design. After this has been completed, the tubes can be reinserted and welded to the manifolds. This procedure can be used in both the creation of the POC and a full scale model.

Actual process

Due to time constraints, the actual manufacturing process deviated from the ideal process. The tubes in our tube array were too close together to be welded, so we needed to press-fit them into the end plate. We also used JB Weld as a sealant for our tube array. We used a drill press (with no power) to make the holes in the carbon foam before the slots were cut. We used the drill press because it allowed us to keep the holes parallel through the foam. We didn't use the power because the dust that would be created is harmful and the foam could be formed by hand. Since the wire EDM would have taken a number of days to cut our big block of foam, and was broken during our time block for cutting, we used a table saw (with no power) to form the channels instead. By choosing to use the table saw, we weren't able to achieve the dimensions we originally calculated; however, the process took considerably less time to complete as compared to the EDM. We used the table saw fence as our base for all of our dimensions. We didn't use power because the blade could cut the foam when turned by hand and no harmful dust was created. Once the channels were created, we pushed the tube array through the holes by hand. Once through, we press-fit the remaining tube array plate and sealed the tubes with JB Weld. Since our sheet aluminum was thicker than expected (due to availability), our bending method was modified. We cut the faces of the end caps and welded each edge together. We were unable to weld our end caps onto the tube array plate, so JB Weld was also used to attach the end caps to the tube array plate.

Testing Plan

Testing will be conducted using the POC as described above and a Lytron straight-finned liquid air heat exchanger which was supplied by our sponsor. Additional information including technical specs of the Lytron radiator can be found in Appendix H on page 50. The same fan will be used to draw air over the heat exchangers to maintain a constant airspeed. The heat drawn off each exchanger can be measured using a turbine flow meter and two thermocouples placed in line of the fluid flow. A thermocouple is placed in line with the fluid flow into the heat exchanger and one placed on the flow out of the exchanger. The flow meter can be placed on either fluid line. A fluid heating element consisting of a large, round bar of copper with a hole in the center for the fluid to flow through with several cartridge heaters placed around it will be

31

used to heat the water. A variable pump will allow us to control the fluid flow rate. Figure 25 below shows a schematic of the setup to be used.

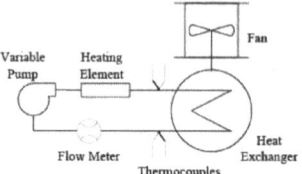

Once the system is set up, either heat exchanger can be placed in the set up for testing. The thermocouples will measure the temperature difference on either side of the heat exchanger and the turbine will measure the volumetric flow rate. From the volumetric flow rate and the density of the fluid, the mass flow rate can be determined. Using the experimental data and the specific heat of the fluid, the heat transferred can be determined using the Eq. 16 below, where m is the mass flow rate, Cp is the specific heat, and ΔT is the temperature difference.

$$Q = \dot{m} C_p (\Delta T) \qquad \text{(Eq. 16)}$$

Testing Results

The proof of concept (POC) radiator was run at four different flow rates (0.25, 0.50, 0.75, and 1 L/min) at an inlet fluid temperature of 80°C. A temperature difference (ΔT) of 19, 16, 9, and 6 °C was recorded for each flow rate, respectively. These results can be seen graphically in Figure 25 below on page 33. These temperature differences translate into a heat transfer of 330, 556, 469, and 417 W, respectively. The benchmark radiator, the Lytron copper, straight-finned, liquid-air heat exchanger performed twice as well under the same conditions. Due to time, only two tests were run on the Lytron but additional results were derived from test data published by the company on their website. The temperature data for both radiators show a general linear trend downwards as flow rate increases though offset from each other by a factor of two with the Lytron above the POC. Heat transfer data for both radiators show a peak heat exchange between 0.25 and 0.75 mL/min, but, again, offset from each other. The results show an overall failure of the POC to perform according to our expectations and calculations. On a positive note, though, the fact that the POC did perform as a heat exchanger and produced actual, viable results is an accomplishment in and of itself. Further discussion of the reasons for the poor performance of the radiator is discussed in following sections.

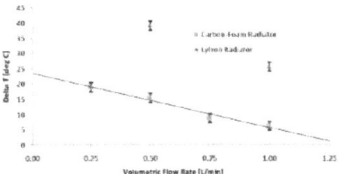

Temperature Change from 80°C Inlet

Figure 25: Temperature change from and inlet of 80°C when run at various flow rates

Discussion for Future Improvements

Analysis of Design

Since our initial tests indicated a lower performance than we expected, there is a need for an analysis of our design and possibly some modifications. Due to the constraints of our manufacturing methods, we were unable to achieve the precise dimensions that were calculated in our engineering analysis. Also, numerous compromises were made during manufacturing. These included using a smaller number of tubes, using tubes with large wall thicknesses (negating our assumption of a thin-walled tube, which was used in the engineering analysis) and using epoxy to seal the manifolds, which caused leaks. An additional flaw in our design was the throttling effect, which trapped the hot air in the channels instead of removing it as originally intended.

Future Work

In order to improve the performance of our carbon foam radiator, several design elements need to be improved. One major improvement is to increase the airflow through the carbon foam. Since air tends to flow over the carbon foam when it is allowed to flow freely, a way of forcing the air into the pores is still necessary, but perhaps a nozzle structure would improve performance. Thinner carbon foam walls would also help to combat this problem, in which case a new manufacturing method must also be developed. The current method of using a table saw is very inaccurate, and a wire EDM is recommended. A new manufacturing method would also be needed for mass production due to the large time requirement of current methods. The tube array layout could also possibly be improved to increase the performance of our radiator. Instead of using the parallel tube setup, a single tube with multiple passes could be beneficial. This should increase the time that the fluid is exposed to the flowing air which would allow for more heat transfer. As mentioned before, the tube thickness could also be reduced, which would allow for less time for the heat to conduct through the tube, and thus increase heat removal.

33

Conclusions

Our task was to design a new concept for an automotive radiator. It was required to reject an increased amount of heat (5%) from current radiator designs while lowering the fluid inlet temperature (10%). A more versatile shape would also be beneficial. We have created a Gantt chart and a QFD. We researched current designs and existing concepts. From this information we created a FAST diagram and a Morphological chart. We generated several concept sketches and evaluated them using the Pugh Chart. Once we chose a final design concept, we went through an engineering analysis of our design to get final dimensions for a carbon foam radiator. Manufacturing and testing of the proof of concept (POC) has been completed. Our design showed great promise in theory but failed to perform to expectations in the lab. Possible reasons for this failure stem from the compressed manufacturing schedule imposed for this project. Due to the lack of time, original plans to use wire EDM for cutting of the carbon foam were scrapped and a table saw was used. The coarse resolution of the table saw forced design changes in the POC and resulted with thicker foam sections than originally anticipated. Other possible design decisions that could account for the concept's underperformance could be a decreased amount of tubes as compared to that originally planned, a large tube wall thickness (which would negate our assumption of a thin-walled structure in the engineering analysis), and using epoxy to seal the end caps (which caused some leakage). These factors may help to explain the failure of the POC for this project. Given additional time and resources further concepts could be built and tested.

Acknowledgements

Bob Coury of the UM undergraduate student machine shop for his assistance and expertise in manufacturing our proof of concept
Professor Albert Shih of UM for his support of our project
Steven White of UM for his guidance and support throughout the project
Professor Katsuo Kurabayashi of UM for his guidance and suggestions for improvement
Professor Claus Borgnakke of UM for his expertise on thermal systems
Thomas Golubic of Koppers, Inc. for his generosity in the donation of carbon foam

References

[1] Mr. Steve White, U of M, Mechanical Engineering PhD student, College of Engineering

[2] Sonntag, Borgnakke, Van Wylen 2003. Fundamentals of Thermodynamics, Sixth Edition. New Jersey: John Wiley & Sons, Inc.

[3] Q. Yu, A.Straatman, and B. Thompson, "Carbon-Foam Finned Tubes in Air-Water Heat Exchangers," *Applied Thermal Engineering*, 26 (2006) pp. 131-143.

[4] C. Harris, M. Despa, and K. Kelly, "Design and Fabrication of a Cross Flow Micro Heat Exchanger," *Journal of Microelectromechanical Systems*, vol. 9, no. 4, pp. 502-508, December 2004.

[5] A. Joardar and A.Jacobi, "Impact of Leading Edge Delta-Wing Vortex Generators on the Thermal Performance of a Flat Tube, Louvered-Fin Compact Heat Exchanger," *International Journal of Heat and Mass Transfer*, 48 (2005) pp. 1480-1493.

[6] Incropera, DeWitt, Bergman, Lavine 2007. Fundamentals of Heat and Mass Transfer, Sixth Edition. New Jersey: John Wiley & Sons, Inc.

[7] Munson, Young, Okiishi 2006. Fundamentals of Fluid Mechanics, Fifth Edition. New Jersey: John Wiley & Sons, Inc.

[8] www.ms.ornl.gov/researchgroups/CMT/FOAM/foams.htm

Biographies

Brandon Fell

Brandon Fell was born in Livonia, Michigan, on May 1st, 1986. He has lived in Northville, Michigan for his entire life and graduated from Northville High School in 2004. He is a senior in his 7th semester at the University of Michigan. Brandon first became interested in mechanical engineering around the age of 12 through his grandfather and uncle, who both worked for Chrysler. He enjoyed building models and watching his family work on cars as a child, and has always been interested in mathematics and the sciences. For the past two summers, Brandon was an Intern at the Detroit Diesel Corporation, working with the Application Engineering Department on various projects. In the future, Brandon would like to work in the automotive industry, with a particular interest in internal combustion engines.

Scott Janowiak

Scott Janowiak was born and raised in Saginaw, MI. When he was about 11 or 12 years old, he started to get an interest in cars. His dad, who is a mechanical engineer for EATON Corporation helped Scott to understand how a car works. Scott has a Jeep Wrangler which he modifies for off-roading and better performance. Last summer he rebuilt his Jeep engine and upgraded several components for better performance.

Scott would like to work in the auto industry when he graduates from U of M. His dream job would be to work on a diesel engine for the new Jeep Wrangler. He also plans on working in the industry for a couple years then going back to school to get his masters degree. Scott is currently a motor gopher for Automobile Magazine. It is a great part-time job because it lets him experience a lot of new cars. He can see which designs are good and which ones can be improved upon. Besides automobiles, he is also very involved in hockey. He plays roller hockey for U of M. He also plays ice hockey in a local men's league. Besides playing, he also coaches youth hockey and teaches learn to skate classes.

Alexander Kazanis

Alexander Kazanis is from Novi, MI and is in his ninth semester at the University. His interest in engineering came at an early age by his fascination in aircraft and from his father, a civil engineer. Aviation has been a great motivator in the pursuance of a mechanical engineering degree and a job in the aviation industry. He currently holds a private pilot license, is working towards an Instrument Rating and works for a refueling company at the Ann Arbor Municipal Airport. Over the last summer, he held an internship at Piper Aircraft Company in Vero Beach, FL. Future plans include working in the aerospace industry, finishing the Instrument Rating and, further down the road, owning his own aircraft.

Jeffrey Martinez

Jeffrey is originally from Grand Rapids, Michigan. He has an interest in Engineering because he likes to know how things work. Also, he has always done well in Math and Science courses. The reason he chose Mechanical Engineering is that he likes the idea that most of it is associated with things that you can see or relate to in the real world. The courses Jeffrey really enjoys are heat transfer type courses because they make the most sense. His future plans are to graduate from the college of Engineering with a Bachelors of Science in Mechanical Engineering in December 2007. After that, he hopes to obtain a stable career. He would like to go into the automotive industry. Some interesting things about Jeffrey are that he is a transfer student from Grand Rapids Community College. Also, before attending college, he was in the United States Marine Corps for four years. While in the Marine Corps., he worked on computers and helped to maintain the base network. He was also given the opportunity to travel while he was in the service. Some of the places that he has been to are Japan, Singapore, Hawaii, the Middle East, and Thailand. Some of Jeffrey's other interests include bodybuilding, cycling, and riding dirt bikes.

Appendix A

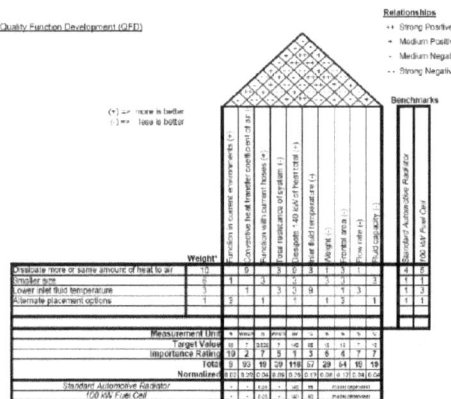

Quality Function Development (QFD)

Relationships
- ++ Strong Positive
- + Medium Positive
- – Medium Negative
- – – Strong Negative

Benchmarks

(+) ==> more is better
(-) ==> less is better

Key:
9 => Strong Relationship
3 => Medium Relationship
1 => Small Relationship
(blank) => Not Related

*Weights are figured on a scale of 1 to 10
(ten being most important)

Appendix B

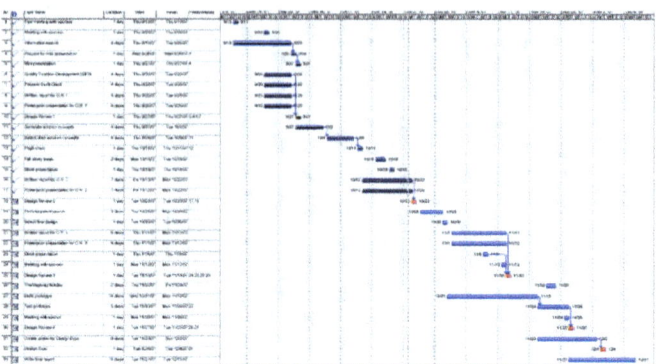

38

Appendix C

Dimensioned Drawing of Tube Cube **All dimensions are in mm**

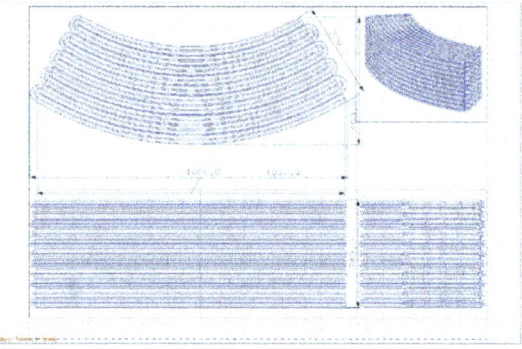

39

Appendix D

Front View of Tube Cube

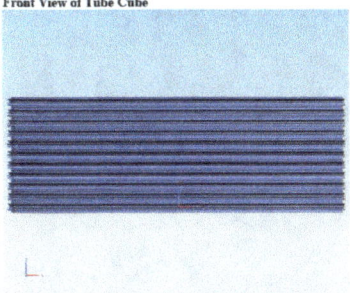

Right View of Tube Cube

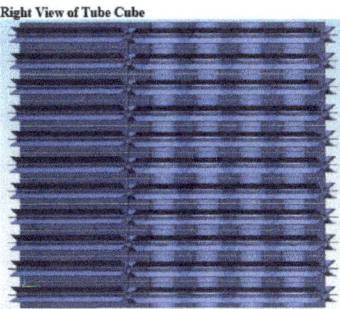

Top View of Tube Cube

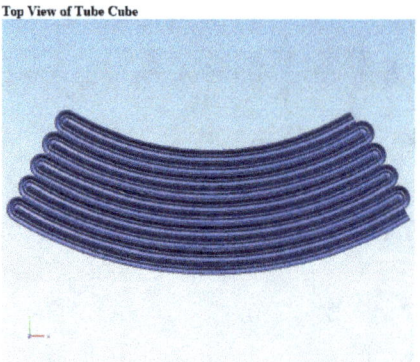

Appendix E

Summary of calculated resistances, initial calculations

Section	Resistance	Value	Maple Name
A	$R_{conv,air-foam\ 1}$	$1/(1.524X+1.016)$	Rcf1
A	$R_{cond-foam\ X}$	$1*10^{-4}/X$	Rkfx
B	$R_{conv,air-tube}$	$.1119/X$	Rcat
B	$R_{conv,liq-tube,1}$	$.0081696/X$	Rclt1
C	$R_{conv,foam\ 2}$	$1/(6.0533Y+2.794)$	Rcf2
C	$R_{cond-foam\ y}$	$.0015625/Y$	Rkfy
D	$R_{K,ls}$	$2.0867*10^{-4}/Y$	Rkls
D	$R_{conv,liq-tube,2}$	$.0081696/Y$	Rclt2

Y(meters)	R_{tot} (K/W)
.005	.08354
.0005	.14664
.00075	.140739
.003	.103298
.002	.1171488
.0029	.104533
.0021	.115599
.00205	.116369

Summary of calculated resistances, corrected calculations

Section	Resistance	Value	Maple Name
A	$R_{conv,air-foam\ 1}$	$1/(1.524X+1.605)$	Rcf1
A	$R_{cond-foam\ X}$	$1.103*10^{-5}/X$	Rkfx
B	$R_{conv,air-tube}$	$.1119/X$	Rcat
B	$R_{conv,liq-tube,1}$	$.0081696/X$	Rclt1
C	$R_{conv,foam\ 2}$	$1/(7.128Y+4.4)$	Rcf2
C	$R_{cond-foam\ y}$	$1.724*10^{-3}/Y$	Rkfy
D	$R_{K,ls}$	$0.00319/Y$	Rkls
D	$R_{conv,liq-tube,2}$	$.0081696/Y$	Rclt2

Y(meters)	R_{tot} (K/W)
.00254	.08649
.01	.05644
.001	.0972
.002	.08995
.0005	.101227

42

Sample Maple™ code, initial calculations

```
> Rclt1:=.0081696/x;
```

$$Rclt1 := \frac{0.0081696}{x}$$

```
> Rcf2:=1/(6.0533*y+2.794);
```

$$Rcf2 := \frac{1}{6.0533y + 2.794}$$

```
> Rcf1:=1/(1.524*x+1.016);
```

$$Rcf1 := \frac{1}{1.524x + 1.016}$$

```
> Rcat:=.1119/x;
```

$$Rcat := \frac{0.1119}{x}$$

```
> Rkls:=2.0867*10^(-4)/y;
```

$$Rkls := \frac{0.000208670000(}{y}$$

```
> Rkfx:=1*10^(-4)/x;
```

$$Rkfx := \frac{1}{10000x}$$

```
> Rkfy:=.0015625/y;
```

$$Rkfy := \frac{0.0015625}{y}$$

```
> Rclt2:=.0081696/y;
```

$$Rclt2 := \frac{0.0081696}{y}$$

```
> R1:=1/(1/Rcf1+1/Rkfx);
```

$$R1 := \frac{1}{10001.524x + 1.016}$$

```
> R2:=1/(1/Rcat+1/Rclt1);
```

$$R2 := \frac{0.007613736033}{x}$$

```
> R3:=R1+R2;
```

$$R3 := \frac{1}{10001.524x + 1.016} + \frac{0.007613736033}{x}$$

```
> R4:=1/(1/Rkls+1/Rclt2);
```

$$R4 := \frac{0.000203472844\}{y}$$

```
> R5:=R4+Rkfy;
```

$$R5 := \frac{0.001765972845}{y}$$

```
> R6:=1/(1/Rcf2+1/R5);
```

$$R6 := \frac{1}{572.3134228y + 2.794}$$

```
> R7:=1/(1/R6+1/R3);
```

$$R7 :=$$
$$1 \Bigg/ \Bigg(572.3134228y + 2.794$$
$$+ \frac{1}{\dfrac{1}{10001.524x + 1.016} + \dfrac{0.007613736033}{x}} \Bigg)$$

```
> simplify(R7);
```

$$\left(3.175000001\,10^8 \left(1.51868038710^{14}\,x + 1.52274720710^{10}\right)\right) \Big/$$
$$\left(2.75958671610^{25}\,y\,x + 2.76697651410^{21}\,y + 1.35356377810^{23}\,x\right.$$
$$\left. + 1.35082143310^{19} + 6.25095250010^{24}\,x^2\right)$$

```
> Rtot=R7/2;
```

$$Rtot = 1 \Bigg/ \Bigg(2 \Bigg(572.3134228y + 2.794$$
$$+ \frac{1}{\dfrac{1}{10001.524x + 1.016} + \dfrac{0.007613736033}{x}} \Bigg) \Bigg)$$

```
> Rclt1:=.0081696/x;
```

$$Rclt1 := 3.21637795$$

```
> Rcf2:=1/(6.0533*y+2.794);
```

$$Rcf2 := 0.356327219$$

```
> Rcf1:=1/(1.524*x+1.016);
```

$$Rcf1 := 0.980516201$$

```
> Rcat:=.1119/x;
```

$$Rcat := 44.0551181$$

```
> Rkls:=2.0867*10^(-4)/y;
```

$$Rkls := 0.101790243$$

```
> Rkfx:=1*10^(-4)/x;
```

$$Rkfx := 0.0393700787$$

```
> Rkfy:=.0015625/y;
```

$$Rkfy := 0.762195121$$

```
> Rclt2:=.0081696/y;
                              Rclt2 := 3.985170073
> R1:=1/(1/Rcf1+1/Rkfx);
                              R1 := 0.0378502984
> R2:=1/(1/Rcat+1/Rclt1);
                              R2 := 2.99753387!
> R3:=R1+R2;
                              R3 := 3.035384170
> R4:=1/(1/Rkls+1/Rclt2);
                              R4 := 0.0992550463
> R5:=R4+Rkfy;
                              R5 := 0.861450168!
> R6:=1/(1/Rcf2+1/R5);
                              R6 := 0.2520642470
> R7:=1/(1/R6+1/R3);
                              R7 := 0.232737287!
> simplify(R7);
                                 0.232737287!
>
> Rtot=R7/2;
                              Rtot = 0.116368643!
> x:=.00254;
                                 x := 0.00254
> y:=.00205;
                                 y := 0.00205
>
```

Sample Maple™ code, corrected calculations

```
>Rclt1:=.0081696/x;
                                 0.0081696
                                 ─────────
                                     x
>Rcf2:=1/(7.128*y+4.4);
                                     1
                                 ───────────
                                 7.128 y + 4.4
>Rcf1:=1/(1.524*x+1.605);
                                     1
                                 ─────────────
                                 1.524 x + 1.605
>Rcat:=.1119/x;
```

45

```
                                      0.1119
                                        x
>Rkls:=-.00319/y;
                                      0.00319
                                        y
>Rkfx:=1.103*10^(-5)/x;
                                  0.00001103000000
                                         x
>Rkfy:=1.724*10^(-4)/y;
                                   0.0001724000000
                                         y
>Rclt2:=-.0081696/y;
                                     0.0081696
                                         y
>R1:=1/(1/Rcfl+1/Rkfx);
                                        1
                                 ─────────────────
                                 90663.35537 x + 1.605
>R2:=1/(1/Rcat+1/Rclt1);
                                  0.007613736033
                                        x
>R3:=R1+R2;
                     1                  0.007613736033
             ───────────────────  +  ──────────────────
             90663.35537 x + 1.605          x
>R4:=1/(1/Rkls+1/Rclt2);
                                  0.002294185007
                                        y
>R5:=R4+Rkfy;
                                  0.002466585007
                                        y
>R6:=1/(1/Rcf2+1/R5);
                                        1
                                 ─────────────────
                                 412.5468269 y + 4.4
>R7:=1/(1/R6+1/R3);
                                               1
               ────────────────────────────────────────────────────────────
                                              1
               412.5468269 y + 4.4 + ─────────────────────────────────────────
                                           1                0.007613736033
                                    ───────────────────  +  ──────────────────
                                    90663.35537 x + 1.605          x
>simplify(R7);
     0.10000000 10^8 (0.6912868557 10^20 x + 0.1222004633 10^15) / (0.2851881988 10^16 y x
          + 0.5041341339 10^25 y + 0.3043267165 10^28 x + 0.5376820387 10^23
          + 0.9066335537 10^28 x^2)
>Rtot=R7/2;
```

$$Rtot = \frac{1}{2} \cfrac{1}{412.5468269\, y + 4.4 + \cfrac{1}{\cfrac{1}{90663.35537\, x + 1.605} + \cfrac{0.007613736033}{x}}}$$

```
> Rclt1:=.0081696/x;
```
3.21637795:

```
> Rcf2:=1/(7.128*y+4.4);
```
0.227088785:

```
> Rcf1:=1/(1.524*x+1.605);
```
0.621553887:

```
> Rcat:=.1119/x;
```
44.0551181:

```
> Rkls:=.00319/y;
```
6.38000000(

```
> Rkfx:=1.103*10^(-5)/x;
```
0.0043425196$

```
> Rkfy:=1.724*10^(-4)/y;
```
0.344800000(

```
> Rclt2:=.0081696/y;
```
16.3392000(

```
> R1:=1/(1/Rcf1+1/Rkfx);
```
0.00431239093

```
> R2:=1/(1/Rcat+1/Rclt1);
```
2.99753387:

```
> R3:=R1+R2;
```
3.00184626:

```
> R4:=1/(1/Rkls+1/Rclt2);
```
4.58837001:

```
> R5:=R4+Rkfy;
```
4.93317001:

```
> R6:=1/(1/Rcf2+1/R5);
```
0.217095233

```
> R7:=1/(1/R6+1/R3);
```
0.202453668·

```
> simplify(R7);
```
0.202453668·

```
> Rtot=R7/2;
```
$Rtot = 0.101226834$:

```
> x:=.00254;
```
0.00254

```
> y = .0005
```
0.0005

Appendix G

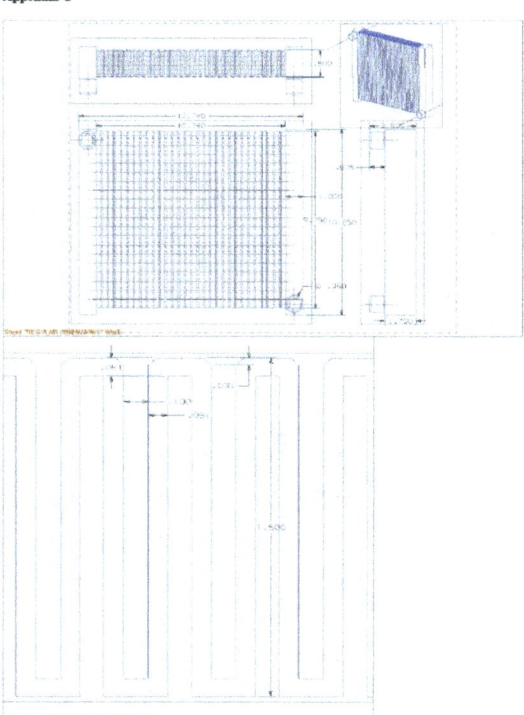

48

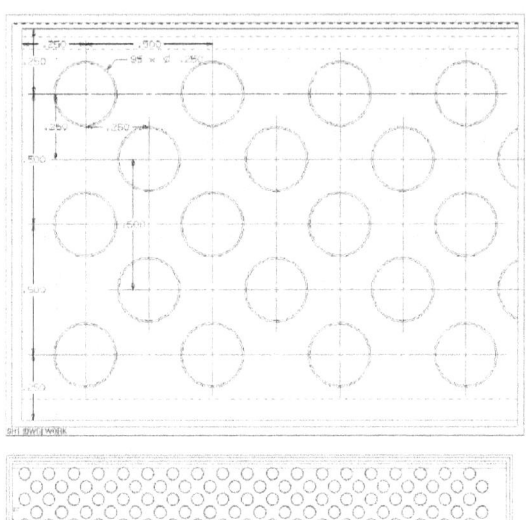

Appendix H

Your Price: $185.21 ea

MSC #: 07440951
Mfr Part #: 6110G1SB

Description:	Tube-Fin Liquid-to-Air Heat Exchangers Heat Exchanger Type: Liquid-to-Air Style: Copper Tubed Recommended Cooling Fluids: Water Fluid Circuit Material: Copper Height: 5.8 In. Width: 7.8 In. Depth: 1.8 In. Connection Type: Straight Number of Fans:
Heat Exchanger Type:	Liquid-to-Air
Style:	Copper Tubed
Recommended Cooling Fluids:	Water
Fluid Circuit Material:	Copper
Maximum Temperature (°F):	400
BTU/Hour:	1140
Height (Decimal Inch):	5.8000"
Width (Decimal Inch):	7.8000"
Depth (Decimal Inch):	1.8000"
Connection Type:	Straight
Connection Size:	3/8" Tube OD
Trade Name:	6000 Series
Fan Type:	76939909
Number of Fans:	1
Manufacturer's Part Number:	6110G1SB
Big Book Page #:	4434